SUPERHUMAN

SUPERHUMAN

Life at the Extremes of
Mental and Physical Ability

ROWAN HOOPER

Little, Brown

LITTLE, BROWN

First published in Great Britain in 2018 by Little, Brown

1 3 5 7 9 10 8 6 4 2

A CIP catalogue record for this book
is available from the British Library.

Hardback ISBN 978-1-4087-0946-7
C format ISBN 978-1-4087-0947-4

Typeset in Perpetua by M Rules
Printed and bound in Great Britain by
Clays Ltd, St Ives plc

Papers used by Little, Brown are from well-managed forests
and other responsible sources.

MIX
Paper from
responsible sources
FSC® C104740

Little, Brown
An imprint of
Little, Brown Book Group
Carmelite House
50 Victoria Embankment
London EC4Y 0DZ

An Hachette UK Company
www.hachette.co.uk

www.littlebrown.co.uk

To Laura

Contents

CONTENTS

INTRODUCTION

A few years ago I found myself at a conference of primatologists. At the drinks reception I enthused – I thought to a sympathetic scientist – about how like us chimps were, and how the differences between us were just a matter of degree. I was very pro-chimp; perhaps I was trying to ingratiate myself among the primatologists. But my stance also reflected the stories I'd been writing, about animals with traits once thought to be uniquely human. There were wild chimps seen using sticks as dolls,[1] chimps using spears to hunt other vertebrates,[2] chimps using their own sign language,[3] chimps waging a war[4] and even what appeared to be chimps practising a proto-religion.[5] My view on these findings, as an evolutionary biologist, was that it is self-evident that we share traits and even behaviours with other animals: we're all related; we share many genes; genes influence behaviour. There's no surprise there. Everything, even what we call 'good' and 'evil', has a basis in evolution, so it's only to be expected that we would find echoes of ourselves in other animals.[6] As a journalist I liked the fact that there was, as I saw it, a lack of uniqueness in humans, as I felt it brought out the similarity between us and other animals, and could even increase empathy between us and them.

1

Wine glass in hand, I blithely asserted that there was nothing unique about us humans. And look at the genetics too, I said, we're practically identical. The primatologist I'd been chatting to smiled an assassin's smile, and said: 'Can chimps build their own LHC then?'

That single remark brought down years of my thinking that we and chimps were so alike. It was just after the Large Hadron Collider at CERN in Switzerland had been used to discover the Higgs boson. The scales fell from my eyes. It's not that I had been overstating what animals were capable of: I had been underestimating humans. It seems ridiculous now, absurd even. The primatologist might as well have asked when a chimp last walked on the moon, or painted *Guernica*. Sure, chimps are wonderful, intelligent animals, but the remarkable thing is not how smart *they* are, but how utterly amazing *we* are. As a biologist, I studied animal behaviour in the field. I marvelled at the solutions that natural selection finds to the problems of making a living and finding a mate. I still do. What I sometimes forgot to appreciate are the marvels of human behaviour and ability.

In some ways this book is an attempt to put myself right on that point. I've set out to meet people at the top end of human potential, across a range of traits. People who are the best in the world at the things we revere, such as intelligence, musical ability, bravery and endurance. We'll also encounter the people who are at the extremes of the things that matter most to us, such as happiness and longevity. It's a celebration of the very best we can be. In meeting them, we'll marvel at the diversity and potential of the human species, we'll try to understand how they personally got to where they are – and we'll deconstruct them. Such people might be superhuman but they're not super-natural. I want to understand how these superpeople do what they do in order to bring them closer to the rest of us. Some of the stardust might rub off on us, and might give us a glimpse of humans in the future. Understanding what lies beneath extreme ability in no way destroys the magic; if anything, it deepens our appreciation and teaches us about our everyday lives. Moreover, we might not be superhuman ourselves but we do have a greater capacity than we realise. We have hidden depths. These traits are the things that humans yearn to be better at and strive to improve.

For most of the characteristics we'll look at, it's fairly easy to decide who is the best in the world, even if my technique is non-scientific. I'm defining the best singers in the world as the ones who can earn a living from their trade; the people with greatest endurance as those who can run the furthest; the longest-lived people in the world – well, they are self-defined. For other traits, such as bravery and intelligence, it's more subjective, but I hope to convince you that I've chosen worthy candidates.

The book is in three parts. Part One, 'Thinking', looks at traits driven by cognitive ability. It takes case studies representing intelligence, memory, language ability and focus – the ability to concentrate the mind. In Part Two, 'Doing', I've picked out bravery, singing and endurance as abilities that humans have taken further than any other animal. Finally, in Part Three, 'Being', I've selected longevity, resilience, sleeping and happiness as traits that at first glance just seem to be part of us, but which some people manage to do at a much higher level. For each characteristic I look at the scientific understanding of how people get to the peak of potential, and the relative importance of nature and nurture – genetics and environment – in each case. There are many clues as to how my superhumans became so good, and lots to learn for the rest of us. The eleven traits and abilities certainly don't capture everything that makes us human, but I think they cast a wide net. The haul has reminded me of the sheer richness of the human species and filled me with fire for our extraordinary potential.

Part I

THINKING

1

INTELLIGENCE

Suppose knowledge could be reduced to a quintessence,
held within a picture, a sign, held within a place which is no
place. Suppose the human skull were to become capacious,
spaces opening inside it, humming chambers like beehives.

Hilary Mantel, *Wolf Hall*

You know it when you see it. I saw it in an orangutan once, a young male in Malaysian Borneo who had been orphaned by deforestation. I was hiking around a protected area of rainforest with a primatologist friend when we came across him.

Because he had been raised in a rehab centre he was well disposed to humans, and, it turned out, especially fond of men. He came bounding over. I was nervous as this juvenile but powerful ape tugged at my clothes and tried to climb up me as if I was a tree. I pushed him away a few times, and he finally sat on his haunches, looked up and held out his hand. I remember taking the hand and feeling it clasp gently and warmly and softly around mine. I caught his eye. In it there was a complex look, a mixture of exasperation, cajoling and hope; he was fed up with me pushing him away, but hoped I would understand that he just wanted to play.

You know intelligence when you see it, and I saw it in him. After that handshake and the look that passed between us, we played for a good hour, which mostly consisted of him climbing up me and me swinging him around. He was basically a monstrously strong hairy orange toddler. He was six years old then and sometimes I wonder what happened to him, and whether he's safe in that protected fragment of rainforest.

It's a special memory for me, but the anecdote exposes several problematic issues with the study of intelligence. Perhaps I was projecting those feelings on to the animal. Many people would say they've seen dogs with the same look in their eyes. Dogs and orangutans might well be intelligent in some sense – but in what sense? How do we measure it?

To study intelligence, we need to be able to define it and measure it, and both things are surprisingly tricky. It's not something like height, which is easy to measure, though crucially intelligence *is* like height in that people have varying amounts of it. Intelligence is complex, multi-faceted, shifting and slippery, and it's the quality we revere above perhaps any other. How strange that we find it hard to agree on a definition. Here's what the American Psychological Association Task Force on Intelligence settled on: 'individuals differ from one another in their ability to understand complex ideas, to adapt efficiently to the environment, to learn from experience, to engage in various forms of reasoning, to overcome obstacles by taking thought'. That's fine, but I want to know about how artists and scientists create and develop new ideas that take us places we've never been.

Intelligence is something we can easily recognise in others, and with IQ (Intelligence Quotient) tests we can measure at least some aspects of it, but giving it a value doesn't tell us what it's *like* to be more intelligent. And what about those people who have never had an IQ test? We'll have a look at IQ later in the chapter, but I want to start – as I'll do throughout the book – by meeting people who exemplify the trait in question. So-and-so might have an IQ of more than 150, but how does that make them *feel*? Where does intelligence come from? What benefits, if any, does it bring? How do people with a surplus of it see the world? Can we load the dice so our children have more of it?

The first person I've decided to meet in this examination is a chess grandmaster. I chose chess because it seems to be a game of pure intellect, or one that is at least highly cerebral. It has also been extensively studied by scientists. It's been said that chess is to cognitive science what the fruit fly *Drosophila*, perhaps the most well-studied organism on Earth, is to genetics.

John Nunn is one of the finest chess players of all time. At his peak, he was in the world top ten. When he was fifteen he went to Oxford to read maths, becoming the youngest undergraduate since Cardinal Wolsey in 1490 (thus handily providing me with a thematic link to someone else we'll meet in this chapter), and went on to take a Ph.D. in algebraic topology, a subject into which I can offer no meaningful insight whatsoever.

Nunn turned chess pro at twenty-six. He was clearly something special, yet while he did scale great heights, he didn't claim the top prize. Commenting on why Nunn, now sixty-one, never became world chess champion, Magnus Carlsen, the highest-ranked chess player in history, said Nunn was *too* clever: 'He has so incredibly much in his head. Simply too much. His enormous powers of understanding and his constant thirst for knowledge distracted him from chess.'

It's fair to say I'm a bit intimidated ahead of meeting John

Nunn. Aware of my hazy understanding of his field of maths, I turn to Wikipedia, which tells me its goal is 'to find algebraic invariants that classify topological spaces up to homeomorphism, though usually most classify up to homotopy equivalence'. I am none the wiser, and possibly even less wise than before. It would make a nice story to have us chatting over a game of chess, but I don't even want to suggest it. There's no false modesty here: it would embarrass him to have to stoop so low. It would be like me suggesting to Usain Bolt that we have a quick run round the park. This is the man who in 1985 beat Alexander Beliavsky from the Soviet Union in a match described as 'Nunn's immortal'. *Chess Informant*, the bible of chess information for players and scholars – a sort of chess *Wisden* – lists the Beliavsky match as the sixth best game ever played, from 1966 (when it started recording matches) to the present day.

We've arranged to meet in a coffee shop in Richmond, south-west London. I get there ten minutes early and secure us a table. Our communication up until now has only been by email, and as such our relationship has been rather formal. I've no idea what he's going to be like in person, but here he is, rocking up in jeans and Converse, with a black motorbike jacket over a hoodie. I hadn't really thought about what he'd look like, but now I've seen him I realise I didn't expect such a groovy grandmaster.

He started playing chess at four. As far as his memory goes, he says, he could've been born playing chess. 'I don't really remember learning to play it.' But it soon became clear that he had an innate talent. How did it become clear? 'Well,' he says mildly, 'when you start winning lots of tournaments it's pretty obvious.'

Immediately it feels we're on to something interesting about intelligence. When Nunn says he had an innate talent at chess, he's saying that genetics played an important part. Of course he had to learn the game, but he claims to have had a natural skill that helped him become good at it. This cuts to the heart of what talent is, and the extent to which expertise in something develops through innate ability and practice. It's something we'll meet repeatedly throughout this book.

There are two schools of thought when it comes to understanding expertise, and they broadly divide along the long-drawn lines of nature or nurture. On the nurture side is Anders Ericsson, a Swedish professor at Florida State University's department of psychology. His work was behind the popular (if now widely criticised) idea that 10,000 hours of practice at anything will make you an expert (we'll come back to this in Chapter 6). Deliberate practice, Ericsson says, can allow anyone to achieve exceptional performance.

I said that the two camps divide along the lines of nature and nurture, but the problem with the whole argument is that the lines shouldn't really exist. Nothing works alone. Genes need an environment to work in and no amount of practice will help if you just don't have the genetic tools in the first place. The argument is really over the relative import of genes and practice.

Zach Hambrick, who runs the Expertise Lab at Michigan State University, could be said to represent the opposing camp to Ericsson. 'Practice is certainly an important factor,' Hambrick tells me, 'but doesn't account for all the differences across people in terms of skill, so other factors have to contribute.' And the factors we're all interested in are the genetic ones.

Look at Magnus Carlsen, the highest-rated chess player in the world by a wide margin. Yet an analysis of the amount of practice he and the next ten highest-ranked grandmasters have put in shows he has practised for significantly *fewer* years than the other players.[1] Does he have talent? In other words, is there a genetic advantage to his ability? 'The answer to this question is so obvious in the chess world that it is not even posed – Carlsen is known as the "Mozart of chess",' say the authors of the analysis, Fernand Gobet of the University of Liverpool and Morgan Ereku of Brunel University.

We'll return to the role of practice in the chapter on music, so for now let's go back to Nunn and dig deeper.

'Magnus Carlsen said of you,' I say to Nunn, 'that you were effectively too clever for your own good, and that's why you never won the world championship.'

'That was very nice of him,' Nunn says.

'Is it true?'

Nunn shrugs. 'Maybe he was right. To be really successful you have to be a monomaniac. To devote your life to it. Some people just can't. They have other interests, they're not happy if they spend their whole life focused on one thing.'

This seems to ignore that Carlsen himself has a life outside chess. Carlsen says his other interests involve chatting with friends on the internet, playing online poker, skiing and playing football.[2] But those enterprises are not the cognitively demanding interests that Nunn had – astronomy, physics and extremely arcane and high-level maths.

'If you're 100 per cent committed it's very painful when things go wrong,' Nunn says, 'because you don't have anything else to fall back on.'

Things will inevitably 'go wrong' in most fields of endeavour, as performance and ability start to decline with age. Nunn coped with that by quitting. Numerous studies have shown that what's called fluid intelligence, which relies on working out abstract problems, and speed of mental processing, both decline in efficiency after about the age of thirty. On the other hand another subset of intelligence, known as crystallised intelligence, which utilises real-world information, maintains its peak level for many years before slowly declining. Nunn seems reluctant to accept that anything about his brain function has changed, and indeed says that his chess ranking has stayed about the same since his professional playing days.

'What changes is you have family and have other priorities,' he says. 'You don't want to focus exclusively on playing chess. But just for age reasons you get tired more quickly. I still feel I can play as strongly as I did but, you know, you play a long tournament, it's exhausting.'

I spoke to Neil Charness, professor of psychology and director of the Institute for Successful Longevity at Florida State University. Charness and his Florida colleague Roy Roring conducted a study of the change in chess ability across lifespan.[3] Using a database of 5011 chess players, they found that the average peak age – the age at which players on average attained their peak rating – was 43.8 years. They also found that age is 'kinder to the more able', meaning that more highly skilled players tend to show milder declines in rating when past their peak age. 'The problems for older players,' Charness says, 'are probably similar to those for any ageing adult, namely that learning rate slows.' Learning rate drops off by half between your twenties and sixties, and as a result there are lots of rising younger players who can out-work you and beat you. While we're on the maudlin topic of cognitive decline, Charness adds that there are probably some age-related changes in motivation. Also, brain efficiency declines in a number of ways, including memory, attention and speed of processing. 'So, although chess for humans is mainly a game drawing on pattern recognition and knowledge, and despite the fact that you can keep learning throughout life, you may have problems retrieving the relevant information in a timely fashion.'

All this seems to tally with Nunn's experience. But I'm interested in how people at the peak of human potential have got there, and I want to find out what it was like when he was younger. Nunn says sure, he felt different to other kids. Was it obvious to him that he had this strength? 'If you keep winning tournaments it's pretty obvious,' he says. 'I won the London under-twelve championship when I was nine.' Did his success make him cocky? He says he was a balanced kid, if slightly solitary, with friends his own age.

What other signs were there, if any, that there was something different about him? 'When I was very young, before I could read, my parents noticed I was looking through all the books in the bookcase, and they said, "What are you doing? You can't read" – and I said I was looking at how many pages there are in each book. I'd worked out how the numbering worked at the bottom of the pages. So they asked me how many pages there were in certain books, and I knew. So I would say my mathematical talent was evident early on.'

Nunn passed A-levels in pure maths and applied maths in his mid-teens. But why go to university so young? 'I wanted to go. I was fourteen. If I didn't go I'd be hanging about for years. Not a good idea really, for a teenager. So I wanted to go, and my parents agreed. It all worked out quite well.'

I wonder if there was a similar discussion in the Wolsey household when young Thomas (later to be Cardinal) Wolsey went up to Oxford to read theology at fourteen. There would be no younger undergraduate for more than five hundred years, until Nunn himself. (Incidentally, there has since been a younger undergraduate. In 1983 Ruth Lawrence went to Oxford at the age of twelve. Her field of study? Algebraic topology.)

What about looking after yourself at that age? Most students can't work the washing machine. 'That was very tricky,' he says, in a way that lets me know that it wasn't tricky at all, 'but I figured it out.'

More challenging was social life. At university Nunn was too young to drink. 'The difference between fifteen and eighteen is quite a lot. A lot of activities don't appeal very much. On the other hand I had mathematician friends and friends at the chess club. By the time I was seventeen it all felt normal.'

Nunn's assessment of his ability and his success mirror the sort of thing Hambrick sees in his research: that expertise is built on innate skill. 'I think my talent in chess and maths was inherent,' says Nunn. 'But in any activity if you want to go to the top you have to have talent and put in the work.'

I started by assuming that chess was a game of pure intellect. Does that mean people who are chess experts are also more intelligent than average? It appeals to common sense that this would be the case, but it's controversial. Chess experts have tended to practise intensively for years, and it's hard to separate that from innate talent or intelligence.

Hambrick, always keen to determine as far as possible the relative roles of practice and skill, looked into the question with his colleague Alexander Burgoyne, also at Michigan. Burgoyne combed thousands of studies of chess skill, and selected nineteen studies containing some 1800 participants that included measures of objective chess ability and cognitive ability, which in practice means the IQ score. Overall, the team found a link between intelligence and chess skill. 'General intelligence and general cognitive abilities correlate moderately with chess,' Hambrick says.

The link was only moderate, perhaps because the top chess players are all of above-average intelligence. There was a stronger link between chess ability and intelligence in younger players and those at lower levels. Hambrick says that might be because highly intelligent people can become good at chess quickly. Average people can become good but need to practise much more. A follow-up study in 2017[4] aimed to test Ericsson's claim that experts become experts not just because they are smarter, but because they have better access to training.[5] Ericsson's idea is that if scientists or musicians have higher IQs than non-experts, it's because people with higher IQs get selected for expert training at universities. The traditional view is that IQ itself predicts who will become an expert. So the arguments differ on the role of training, with Ericsson emphasising the training. Hambrick's team tested this by comparing chess players with others who don't play chess. The rationale is that in chess there isn't the selection process we see among scientists and musicians trying to gain a position in some academy, so if Ericsson is right then there should be no difference in IQ between chess players and non-chess players. But there was a difference: chess players outperformed non-chess players in cognitive tasks, meaning that training alone can't explain expert performance.

So what innate skills might these be? For chess Nunn says you need a power of visualisation to see what the options are going to be in four or five moves. You need a good memory. A good ability to calculate. You need skills of pattern recognition. Starting young is important in chess. Studies have found that even after controlling for amount of practice, the younger you start, the higher the ranking you are likely to attain.

According to Charness, pattern perception plays a critical role in optimising the search process that chess players use when trying to find the best move, 'So if there are any innate or genetic differences, I would be looking for someone having slightly more fluent pattern abstraction processes.' Not that this means there is such a thing as a chess-playing gene. 'It is hard to see there being evolutionary pressures for expert chess playing, despite the occasional bumper sticker you see that proclaims "chess players mate better".'

All this has helped enlighten me about the components of intelligence that chess players seem to have at a high level, and I understand better how chess skill develops from a young age and tails off with time. But these things don't convey how it feels to play, and they don't tell me what it is about Nunn that got him into the world top ten. So after our chat I watch a narrated reconstruction on YouTube of 'Nunn's immortal': the fabled game against Soviet grandmaster Beliavsky. The reason the game is revered is that Nunn discovers an extraordinary response to Beliavsky's white pieces (aligned, if this means anything to you, in the Sämisch Variation of the King's Indian) that suggests an apparently strong attack for white. Seeing black apparently blunder, white makes the attack. Black then sacrifices one of its knights. A piece down, it would seem the game is up, but black manages to contrive a net from which white cannot escape. Watching the reconstructed game, I think I can grasp the beauty people see in it. Nunn glimpsed something that was possible in the vast gamespace of potential outcomes, but that was concealed from everyone else, even other grandmasters who afterwards analysed the game. This is how chess can achieve the quality we normally associate with art. The Beliavsky–Nunn game represents an idea that hadn't been thought before.

'I was aware during the game that it could be something special, a once-in-a-lifetime opportunity, and I was really motivated not to spoil it by inaccurate play,' Nunn says. 'In the event, everything went well and afterwards I was delighted with the game.'

I ask Nunn if he's ever had an IQ test. He has, when he was young. But he doesn't want to tell me what the result was. 'It was rather high, I think it was unrealistic.' I wheedle the score out of him, but only by promising that I won't publish it. But I can report that yes, it is 'rather high', if by 'rather high' you mean snow-capped and exalted mountain peaks far beyond the troughs, bogs and modest hills of the vast majority of the human species. I goggle at him, mouth flapping open. 'So what,' he says. 'It doesn't mean anything, I suspect.'

Let's make a short diversion into IQ.

When he says his IQ score doesn't mean anything, Nunn is perhaps allowing his modesty as to its value to influence his appraisal of the worth of IQ testing. It's significant, however, that he denies the value of IQ testing, saying that it merely tests how good you are at doing IQ tests. 'This is a bit like a multi-billionaire saying "Oh well, you know, I get along, but money isn't everything",' says Stuart Ritchie, an intelligence researcher at the University of Edinburgh's department of psychology. You do however hear this criticism a lot, so let's take a look at it. IQ scores correlate closely with a wide variety of life outcomes. 'If we were able to take a step back and look at the whole range of people, with much lower and much higher IQs than ourselves, we would start to notice patterns in how it relates to education, health, job success, longevity and so on,' Ritchie says. 'IQ measures something that seems to be pretty important in life, not just how good you are at doing the tests.'

Here's a powerful illustration of that. In 1947, around 94 per cent of the Scottish population born in 1936 completed an IQ test. The test has high lifetime stability, meaning the score the children received at eleven will correlate well with their intelligence scores later in life. In 2017, researchers managed to follow up more than 65,000 of these children. They found that the higher the value of childhood intelligence, the lower the risk of death from a variety of causes.[6] This included death from respiratory disease, heart disease and stroke, and also from dementia and suicide. Socioeconomic status had only a modest impact on the likelihood of dying.

I've never had an IQ test. I'd be nervous of having one because I worry that I wouldn't get a decent score. Actually this fear is misplaced. Findings about IQ such as those discussed in the Scottish study are always based on IQ scores averaged across populations. For individual people, a high or low score is not a good predictor of success of failure.[7] But the more important concern for many people is that it feels like we can potentially order people in merit by their IQ score, when of course there is much more to human life. This is the 'I won't be reduced to a number' argument. Here are two slam-dunk replies to that. First, says Ritchie, 'Nobody ever said it was a single number.' No one claims your IQ does represent your life. IQ is a convenient summary number, but researchers build statistical models of a whole hierarchy of cognitive abilities, going from specific to general, and look at how these relate to different life outcomes.

Second, if we refuse to reduce complex things to numbers, we can't investigate them scientifically – this relates to everything from psychological testing to climate science. 'We're all crushingly aware of how complex intelligence is,' says Ritchie, 'and these numbers and models are just a first step along the way to really understanding why some people are smarter than others.'

The IQ test may be the most controversial measure in science, but psychology has no more rigorously designed and sophisticated assessment. 'All tests and surveys have problems,' says Dana Joseph, a psychologist at the department of management, University of Central Florida, 'but intelligence tests have been studied more than perhaps any other type of test, so these problems have been minimised as much as possible.' IQ tests don't cover *all* forms of intelligence, but no test covers 100 per cent of the construct it intends to measure, and IQ has proved robust across a range of measures. 'We have evidence suggesting that many intelligence tests capture intelligence fairly well,' she says.

A proper IQ test measures a range of different abilities – memory, reasoning (both verbal and abstract), general knowledge, speed of brain processing and spatial awareness. The results for each subset of tests are crunched together to give a final score, and the average for the population is set by the testing companies as 100. Many of us might say something like 'Oh, I'm great with words but hopeless with maths' or 'I'm good at reasoning but I've got a rubbish memory' – we feel we have strengths in certain areas but not in others. But here's the odd and powerful thing about IQ: people good in one subset of the test tend also to be good in others. All the measures come together to give the 'general factor of intelligence' that researchers call g.

IQ isn't the same as the general factor of intelligence, but because it samples lots of different aspects of intelligence it correlates closely. A vast amount of work over many years has shown that people with high scores in intelligence tests do better when they are at school and at work, and even have better health outcomes. We also know that there is a genetic influence on intelligence. There are many genes that each have a small influence on intelligence, which is what we've come to expect from complex traits. When the genetic effect has been measured, it turns out to be smaller for IQ than it is for g. In other words, cultural, educational and social factors seem to have more influence on IQ, whereas g is more biological. When you follow people with high IQ throughout their lives, they tend to be extremely successful: they are more likely to get important and powerful jobs, and become influential in many different spheres of life, from art to music to politics to science. They tend to be healthier because they make better decisions. Incidentally, you are more likely to match your spouse for intelligence than for other traits. The correlation for IQ between spouses is about 40 per cent; it is about 10 per cent for personality and 20 per cent for height and weight.[8]

Deep criticisms remain. One of the most troublesome is that any differences in IQ found between different groups of people – say, African Americans and white Americans – could be used to suggest there are real differences in intelligence. More likely is that if any such differences exist they reflect socioeconomic and cultural differences, such as the experience of racism and poverty, rather than anything genetic.

Another criticism is that IQ ignores emotional intelligence, that is, the ability to understand what others are thinking and feeling. (It's worth mentioning here, as the journalist Adam Grant points out, that important historical figures who had a powerful command of emotional intelligence include Martin Luther King Jr . . . and Adolf Hitler.)[9]

A measure of emotional intelligence is thought to be useful in assessing candidates for jobs involving interactions with people, though work by Dana Joseph and her colleagues suggests that regular IQ tests are better predictors of performance.[10] It seems that the IQ test does the job better than other tests designed with different forms of intelligence in mind.

In summary: yes, there are issues, but IQ is the best measure of intelligence that we have and correlates strikingly with a range of lifetime outcomes. Having said all that, I'm well aware

21

that intelligence is broad, complex and rich, and manifests in any number of different ways. I still want to investigate people with different forms of intelligence, so I've arranged to meet one of our greatest living writers.

Before I meet Hilary Mantel I try to get into character. I walk a short stretch of the Thames in west London, from the Ship pub in Mortlake along a wall that traces the same boundary as did, five hundred years ago, the wall of Thomas Cromwell's manor. Henry VIII made his chief minister Lord of the Manor of Mortlake in 1536. I imagine turning a time-travel dial that would rewind the clock 481 years so I could see Cromwell stepping off the barge from the city, perhaps with a small entourage of staff, after advising the King in court. I try to imagine the area without the modern houses and the pub and the old brewery and Chiswick Bridge. All these things fall away rapidly and there are still hundreds of years left to rewind. Only the shape of the river is the same now as it was in Cromwell's time. Perhaps there are a few old oaks around that were young half a millennium ago. I can't really picture the scene – not like Hilary Mantel, who has pictured it, researched it, imagined it, practically lived it.

Mantel was already a successful novelist before she started writing about Thomas Cromwell. (He, incidentally, must have been one of the most intelligent people in the Tudor period; certainly he was one of the most influential.) But the first two books in Mantel's Cromwell trilogy, *Wolf Hall* and *Bring up the Bodies*, both won her the Man Booker Prize (she's the first woman to win the award twice) and propelled her to international acclaim. *Time* magazine named her in 2013 as one of the hundred most influential people in the world, and she's widely considered one of the world's finest living writers. The depth of her characterisation and the intelligence of her writing is remarked upon by almost every reviewer. If anyone is a candidate for inclusion in this chapter, it's her.

'It's vaguely mock Tudor,' Mantel says when giving me directions to her flat – how appropriate that she lives somewhere that is a modern take on a Tudor style.

It's no surprise that the first thing she identifies as a special skill is her way with words. 'The family story,' she says, 'is that I didn't speak and didn't speak and didn't speak, until I was about two and a half – and then started talking like an adult.'

She says when she thinks back to her early memories they don't feel like what we think of as a child's memory. It's as if the young Hilary experienced an accelerated mental maturation. 'It's almost as if there was a much older person sitting inside me.'

She thinks her linguistic fluency is an innate characteristic. As a young girl she'd sit for days in the company of her grandmother and gran's sister, who would talk constantly, almost ritualistically, perhaps varying only slightly what they'd said the day before, but Hilary would be soaking it all up. By school age she'd acquired an enormous vocabulary. She had been the only small child amid a mass of adults and had become minutely attuned to the rhythms of their language. 'That's where my weird vocabulary came from when I went to school. When I wasn't speaking I think I was working it out, and only speaking when I was good and ready.'

This knack for language is something untrained in the sense that neither Mantel nor other members of her family studied it; her mind seems like it was ready to use correctly everything it heard. 'We're all good talkers. Even my mother, who left school at fourteen, never makes a mistake in syntax or grammar. She can carry the longest sentence – it may be nonsense in context but it's perfect in form. And I think of that as something that must be inherent.'

We've seen already how IQ accurately reflects intelligence, and how it correlates with many aspects of our lives. That in itself has been controversial, but it's less so than the idea that some components of intelligence are inherent. To investigate this, I go to see Robert Plomin, a Chicago native long since transplanted to Britain, where he is professor of behavioural genetics at King's College London. Plomin launched and runs the Twins Early Development Study (TEDS), the UK's largest study of twins, involving more than 15,000 identical and fraternal twins who have been assessed from early childhood to (currently) the age of twenty-one.[11] Because twins share their genes and almost the same environment, it's possible to examine a trait such as intelligence – but also many other traits, like obesity – and determine how much is influenced by genetics and how much by your environment. Plomin also looks at twins that are raised apart, as well as adopted children,[12] and his conclusions are robust, clear-cut and somewhat shocking. Most of the child's academic achievement until the age of sixteen is explained by genetics. It doesn't matter what kind of school the kid goes to, or the kind of home it is reared in – the IQ of the separated child correlates very closely with that of its sibling, or its birth mother. It doesn't correlate at all with the IQ of its foster-sibling or adopted mother. 'All the different approaches: adoption studies, identical twins, twins reared together and apart, they have different assumptions but they all converge exactly on the same conclusion. Genetics is much more significant an influence than school or home environment. I don't see how people can deny the data,' he says.

In a study of 360,000 sibling pairs and 9000 twin pairs,

Plomin and colleagues found that high intelligence, of the sort we're exploring in this chapter, is familial and heritable: almost 60 per cent of the differences in high intelligence are genetic.[13] This supports Mantel's intuition that a substantial part of her intelligence is inherited.

There are no 'intelligence genes', however. Or rather, there are, but there are thousands of them, each with a tiny effect on intelligence, and with effects on other traits. In 2017, a genetic analysis of 78,308 people found fifty-two genes that together explain less than 5 per cent of the variation in intelligence between people.[14] If it's confusing that 60 per cent of intelligence is genetic but the study only found 5 per cent, it's because specific genes are hard to find. That 5 per cent refers to specific genes, and scientists are working on finding the other 55 per cent.

Mantel doesn't call herself a perfectionist, but she cares passionately about getting things right. 'It seems I was always preparing to be a writer. I was always finding the exact words for things, nothing vague would do.' (My response to that was that it also seems like good training for being a scientist.)

By her own reckoning, a good memory and an extreme verbal fluency are her outstanding qualities (and we'll look into these in depth in Chapters 2 and 4). Sometimes she slips into talking about herself in the second person, and the effect is a little like reading her Thomas Cromwell books, which often use the pronoun 'he' to refer to Cromwell. 'You perceive your general intelligence wouldn't come out that high on a test,' she says, 'but what I've got that an IQ test wouldn't tell you is a pretty good memory, and a capacity to mince industrial quantities of data.'

As it happens, a proper IQ test does measure memory, but perhaps not the long-term memory Mantel is talking about. In any case Mantel turns out to be one of those people we've just considered in the discussion on IQ: she thinks she's good with words but bad with numbers. 'I always thought the world was overrating me because I was so verbally fluent. And I could bullshit my way through an exam paper. With mathematics it's a complete blank. I'm not innumerate. But I remember sitting through the first lesson in calculus and it might've been given in Russian for all I understood.'

This is harsh: calculus is not easy even in your native language. I mention that although you often hear people saying things like 'I'm good with words but poor with numbers', a lot of intelligence research has shown that clever people are clever across a range of different kinds of intelligence. Perhaps she just didn't get the nudge she needed to be good at mathematics. She considers this and revises her opinion. 'I had this hobby of working out immensely complex multiplications. I wasn't against numbers. But when it became conceptual rather than mechanical there was something I wasn't getting.' It's a classic case of how a slight 'tilt', as psychologists call it, towards one type of ability, can influence your life's trajectory.[15] In Mantel's case, of course, she tilted to becoming a novelist.

Extremely intelligent children are often lonely, basically because they are bored with kids their own age. As a small child, Mantel says she formed a habit of ignoring people who asked stupid questions. She perceived a lot of school as a waste of time. She found it difficult to make friends. 'I think people thought I was crippled by shyness. I was always on the outside of a group. I would be the person who would think of the game, and everyone else would go off and play it. It's typical of what writers say, they're watching everyone else.'

She says she was frustrated, rather than unhappy. 'But then I had a feeling of biding my time. I was always trying to get somewhere. I had boundless but completely unformulated ambition. So whatever was happening I regarded it as only temporary.'

I think this is getting at something that seems to be key to the way intelligent, or at least successful people function: ambition, or drive. You had a drive, I say.

'That's right. To put your mark on things. It wasn't "I will leave this small town and I will show them", it was more that – you know those experiments they do with children with marshmallows about deferring gratification? I'd've been great at that.'

The marshmallow test was devised by Walter Mischel at Stanford University in the 1960s. Children were presented with a treat, and told they could either have it now, or they could wait fifteen minutes and have two. About a third of the children managed to hold off eating the marshmallow and reap the reward. Several years later, to his surprise, Mischel found that the kids who had shown self-control ended up doing better both socially and academically. The result has been repeated in a number of different ways, and forty years later, when the kids had grown up, researchers tracked them down and found there were differences in the adults' brain networks according to their powers of self-control. Kids, now adults, with better will power had predictably different patterns of neural activity.[16]

Another study found that people with better self-control were more intelligent, and brain scans suggested there was more activity in the anterior prefrontal cortex.[17] This is one of the last parts of the brain to mature, and listening to Hilary Mantel's story, it's tempting to suppose that her brain really did mature more rapidly. She imagined that there was an older person sitting inside her, but I wonder if her intuition may be tapping into something real.

'When I think back to the judgments I made of people as a child they often seem more like adult perceptions,' she says. 'I think I was just amused by people. And if you're looking for the next mad or quirky thing they might do, or you're looking for them to manifest themselves in some extreme way – then you're not so frightened of them, as children can sometimes be. You've detached yourself.'

She checks herself and mentions that she's just hit upon something she hadn't known before, about waiting for people to manifest themselves. 'You have an idea about someone and then you wait for them to prove it to you in their behaviour. I think that's how I operated. As if there's a play to be put on and I was sitting waiting. Not that it wasn't painful. No child likes to feel an outsider. But I don't think I was damaged by it.'

Kristine Walhovd is a professor of cognitive neuropsychology at the University of Oslo, Norway, and one focus of her research is looking at how the brain, and cognitive ability, changes with age. I ask her about how evidently highly intelligent people such as Hilary Mantel might feel different, and how this might be explained. Walhovd stops me getting carried away:

'We are all unique, so when Mantel comments that she felt different, that is of course true. We all are. I bet if being honest, most people, like Mantel, would say that they felt different to "the others" at some point growing up. Wouldn't you? And we are all right, we are different to everyone else, but so are "the others".'

Well, that's good. It's my aim after all to demystify people at the extremes. Mantel is not *uniquely* different, of course she isn't. Nor was her extreme verbal ability an extreme sign of the difference in maturation rate of boys and girls: Walhovd says the sex difference is overrated.

If there are qualities that Mantel has had since she was a girl, they are a certainty of herself as an individual, determination and focus: 'I've always had a strong and definite sense of self that didn't depend on what other people said or how they viewed me. I see myself as being obdurate, like a stone, inside.'

All her projects, since she was quite young, have been big and long term. She knows that something won't pay off for five years but then it will. But why was she deferring gratification? What was her marshmallow?

'It's that when you set your intention, your intention will keep moving ahead of you as you move towards it. And by the end you will have done something bigger and better than you could imagine, and you will have changed and you will have more capacity. So it's always just keeping ahead of you and stretching what you can do.'

If challenging and developing herself was and is her goal, she identifies the driver of this development. 'Curiosity is the basic thing, I think. You really have to be asking yourself every morning, what could happen today? I can see that if you consistently maintain that mindset it will keep your mind lively into old age.' She pauses before continuing: 'Your body's different of course.'

Mantel has suffered poor health for many years, and the lack of physical stamina is one reason, she says, that she didn't go into politics. She picks up again. 'And I perceive that a happy life is one where even way into middle age you're still finding new capacities.'

We'll come back to what constitutes happiness at the end of the book. For now, I want to ask Mantel what I think of as the Atticus Finch question. One of her strengths is her ability to understand the human condition from different viewpoints; it's her knack of getting inside a character's skin. How does she do it? She says she thinks of writing as a performance where you have to play all the roles. 'If it's a character who's plucked out of your head then I perceive them to be in some way aspects of your self, an exploration of unused sides of yourself, like if you were a man what kind of a man would you be. So you're playing out your unlived lives.'

Mantel says that with her main characters she can't see through their eyes, only just to the side. You try to give the main characters free will, she says – I think of Cromwell, and I'm sure she does too – and they feel indefinite around the edges. I didn't grasp what she meant at first, then later on I think she is not omnipotent about them because her main characters are so deeply realised that they almost become their own people.

Her creativity, as she sees it, comes from self-examination. 'I'm keenly noting what passes inside my head and am not so focused on what's happening in the real world,' she says, which leads her to a defence of the pen-chewing, mind-wandering, drifting writer: 'People misunderstand what they call daydreaming.' She then intuitively states what psychologists have only just discovered, using brain scanning.

'For a writer daydreaming is extremely purposive. It's directed. And what you see in your inner cinema reel will get translated into words, or stored and you'll rerun it, but it's not at all a hazy activity.' Paul Seli at Harvard University has shown that mind-wandering is indeed linked to purposeful thought;[18] other studies have indicated that daydreaming can boost creativity and aid problem-solving.[19]

'There's a lot of conflict in myself because I really like to know the facts,' she says, and again I think of the similarities with science. 'A lot of creativity is about tolerating ambivalence. All the time you're dealing in layers or shades of meaning.'

*

Let's turn now to someone who also likes to know the facts, but has a superhuman *intolerance* of ambiguity.

Paul Nurse, who has a string of letters after his name as long as a bus, won the Nobel Prize for physiology or medicine in 2001, for his part in discovering how the cell cycle works. He was driven, he says, by the desire to understand what is the fundamental difference between life and death, and he realised the way to do it was to look at how cells divide.

Working with yeast, he figured out the genes and proteins involved in the regulation of cell division. He also showed that in humans these were the same genes, and that when the process goes wrong it can lead to cancer. He was knighted in 1999, he became president of Rockefeller University in New York, then president of the UK's academy of sciences, the Royal Society, and is currently director of the Francis Crick Institute in London, the biggest biomedical lab in Europe. Once when I was to introduce him at a science festival, I asked how I should refer to him, and he said 'The blob' (obviously I ignored this instruction). He's one of the most influential scientists of our time.

He also has an unusual family pedigree. Living in New York after his Nobel Prize, he decided to apply for a green card, but his application was rejected because the birth certificate he'd submitted was a short version, which didn't show his parents' names. When he retrieved the long form from the UK register office, he found that his father's name was unknown, and under 'mother's name' was the name of the woman he'd always thought was his sister. What had happened was Paul's mother became pregnant by an unknown man, and gave birth to Paul in Norfolk, to hide the illegitimacy. He was brought up by his maternal grandparents, who pretended to be his parents for the rest of their lives. The woman eighteen years his senior, who he thought was his sister but was actually his mother, had died by then. His 'brothers' were his uncles. His father he never knew. If the story was in a soap opera you wouldn't believe it.

Before this intrigue came to light, Paul was raised essentially as an only child, his purported siblings being so much older than he was. He spent a lot of time alone, and on the long walk to and from school an innate and powerful curiosity became apparent. Like Hilary Mantel, he cites curiosity as the central component of his intellect, although in Nurse's case it is supplemented by logic and experimentation, and it is these things that make him a scientist.

I say innate curiosity, but Nurse, when I sit down with him at the Crick Institute, is careful not to assume that his curiosity is a genetic trait: 'There's no doubt that I somehow developed an almost unhealthy curiosity about the natural world around me, and it may have been because I wasn't distracted by too many of the family things.' The time he spent alone made him observe nature. He became interested in astronomy and natural history and in trying to work out how the world worked, and that encouraged a real curiosity about the natural world. It's a curiosity he still has.

He says that, on the whole, most of our higher-faculty characteristics such as intelligence come down to about 50 per cent genetics, 50 per cent environment, sometimes a little more or less. 'As a geneticist I'm fully aware of the impact genes can have. My environmental component was not academic and intellectual so I have rationalised in my head how my environment was encouraging for where I ended up.' Before he discovered that no one knew who his father was, he'd sometimes wondered why he was the intellectual outlier in the family. He didn't have books at home, or parents who encouraged reading or curiosity. But his self-starting curiosity latched on to the stars in the sky and the moths in the hedgerows on the walk to school. I'm reminded of what Robert Plomin said to me. After seeing what happens with twins and adopted children reared in unstimulating environments or sent to poor schools, Plomin knows that ability will out: 'At the higher end of ability, you'd almost have to lock kids in closets to keep them from getting ahead.'

Scientists are supposed to pride themselves on their objectivity and adherence to logic and evidence, but in practice it's all too easy for some to jump to conclusions and cling to a favoured hypothesis based on less than watertight data. Nurse genuinely seems to embody the 'proper' scientific method. He is brutal with ideas. 'I've never felt that having clever ideas by itself is worth very much, because ideas are cheap. They're only worth something when you test them to destruction.'

Nurse identifies the desire to test something to destruction as a component of intelligence. 'There may be something to do with pride there. If I have an idea and have observations to support it, rather than get that out there I go around and look at it in different ways and try and destroy it. And only if it survives do I begin to talk about it. So there's something about destroying your own ideas. It means that when it does survive you speak with great confidence.'

When Nurse arrived at grammar school, and then university, he was surrounded by students from more enriched backgrounds. Kids who'd had books at home, and intellectual encouragement. It made him feel he needed to learn about the world because other people knew much more. So he started taking the *Times Literary Supplement* and the *London Review of Books*, to keep his education broad, and he still does. 'It might've been driven by feeling a bit inadequate but it was more that there are many things out there in the world that I wanted to know about. So it's back to this intense curiosity.'

His curiosity and desire to discover drew him to the life sciences. In the physical sciences, the questions seemed too big and difficult to address and he felt too small a cog. Whereas in biology the problems are things you can observe. He would count spiders' webs in the back garden and wonder about the corresponding distribution of flies. 'I couldn't study the structure of the atom in my garden.'

I mention that he could do it his head, but he says he wasn't in his head, he was always observing and relating things to the world. The idea of internalising problems and thinking about them mathematically was alien, completely outside what he'd been exposed to in his non-academic family. It reminds me of what Hilary Mantel said, that when mathematics at school became conceptual, she stopped keeping up. She too was raised in a non-academic family ('I was at the worst school in Derbyshire at the time'), although outside school she was immersed in books. And it reminds me of the psychology of 'tilt': Mantel's language skills tilted her towards writing; Nurse's curiosity about the natural world tilted him to biology.

Paul Nurse's children, of course, have grown up in an academic, intellectual family. One of his daughters is a high-energy physicist. 'She's next generation on and was brought up in a different way and obviously wasn't daunted by this, but I was,' he says.

At school, as throughout his life, Nurse was quite often 'off piste' – prone to following imaginative leads, taking off on flights of fancy. 'If I was off piste in an interesting way my marks went up, and if not they went down.' And as he grew more mature, going off piste become profitable. 'I'm pretty imaginative when thinking about solutions, but I'm always being dragged back to experiment and observation and testing. Which may even have its roots in that I liked looking at the world.'

Mindful of some of the subjects I'll be investigating – memory and languages – I mention that these are supposed to be things that intelligent people are good at, but he's dumped them. 'They're related. You need to be able to remember that *chien* is dog,' he says. 'But there's another problem – my grandparents had strong Norfolk accents.' He puts on a country bumpkin accent. 'I was living in London where they made fun of me talking like this.'

His inability to pronounce words in a 'normal' way (he means without a strong regional accent) hindered him when it came to languages, he says, and he's no good at mimicking. His memory is a problem, but putting things together in unusual ways is a real strength. 'It's making connections from different parts of the intellectual universe that most people have trouble connecting. It's being voracious across the board, but coupled with logic. I put things together and then apply logic, and if that fails, dismiss it.'

There is a family element to this, it seems. He came from a humble background, and doesn't want anyone else kicking his ideas – perhaps then they would dismiss him as a bumpkin.

Nurse is sixty-eight now. Cognitive powers, especially speed of thinking and the ability to deduce things without prior knowledge, decline from your thirties, but crystallised intelligence – as we've seen, this is the sort you draw on when you need knowledge of the real world, facts and figures, experience – maintains its level for some decades.[20] I ask Nurse what's changed for him.

'I know I'm not such a good thinker as when I was thirty.'

One of the things that's gone is the ability to concentrate for hours on something. 'I saw this in my physics daughter: like a laser on a problem until it vaporises. I think I'm too distractible now to get that complete focus.'

He can compensate to some extent. 'The thought processes you apply to a problem, I've been through before so I know what to do.' Plus a form of cunning sometimes develops, which can help. His raw intellectual endeavour may be weaker but Nurse has experience — some of which he imparts as I leave. First, the railway-track metaphor of ideas.

'When you have an idea to explain a phenomenon, it's actually difficult to think of another idea, and you start to run down rails. You have to find ways to jump the tracks.'

So he has tricks that enable him to do this. Don't work too hard. 'If you're constantly thinking of the problem you're on the same track.'

Read things that aren't necessarily relevant, and do other things that force a change of tack. Nurse is a pilot, and flying forces his brain to switch tracks. 'You think of nothing else while you're up there other than staying alive. And when you come back down again, literally, after gliding in the Alps, you look at everything in a fresh way because you've cleared your brain out.'

Finally, he has an observation on failure. 'You cannot help but fail. Failure happens all the time.' You have to accept this, learn from it, and, crucially, move on. He says a lot of his supervisory advice is about stopping people getting depressed when they fail, and about helping them develop a psychological approach to cope with failure. You have to have motivation to succeed, but you have to make sure your motivation is solid. 'If you want to be famous, you're probably never going to be. There're different motivations. Curiosity drives me.'

2

MEMORY

Memorie, the greatest gift of nature,
and most necessary of all others for this life.

Pliny the Elder, 1st century AD

I construct my memories with my present.
I am lost, abandoned in the present.
I try in vain to rejoin the past: I cannot escape.

Jean-Paul Sartre, *Nausea* (1938)

I'm more than my brain but my memories are what makes
me me, so if I don't remember then who am I? . . . I don't
know when to say goodbye.

Nicola Wilson, *Plaques and Tangles* (2015)

Ireneo Funes was bucked off a horse and knocked out. When
he regained consciousness his body was broken but his memory
had become perfect – superhuman, you might say. He could
recall every dream and every daydream he'd ever had. 'He
knew the forms of the clouds in the southern sky on the morning of April 30, 1882, and could compare them in his memory
with the veins in the marbled binding of a book he had seen

only once, or with the feathers of spray lifted by an oar on the Rio Negro on the eve of the Battle of Quebracho.' Funes, his memory now limitless, constructed a madcap numbering system, whereby numbers were assigned words. So the number 7013 was 'Maximo Perez', and 7014 was 'the railroad'. The system went beyond 24,000 in this way.

Funes, of course, is fictional, a product of the extraordinary mind of the Argentinian writer Jorge Luis Borges. In his short story 'Funes, His Memory', Borges observes that Funes' method of numbering, and his other similarly confusing mental catalogues, are pointless, and intelligible to no one but himself, but they at least show us what's going on in his mind. 'They allow us to glimpse, or to infer, the dizzying world that Funes lives in.'

This is the kind of world we're going to explore in this chapter. We'll meet people who approach Ireneo Funes in the magnitude and accuracy of their memory, but are all the more startling for being real. And we'll go the other way, and see how for all of us, the faculty of memory is delicate, malleable and utterly unreliable: and that this is a good thing. Memory is just as strange as, and arguably more mysterious than, the richest and weirdest Borges story. Bear in mind as we go that there are two meanings to the phrase 'she has an amazing memory'. The first is that she has the *ability* to store a lot of information; the second is that the *content* is remarkable. For the most part we'll be focusing on the cognitive ability.[1]

Let's start with a number. It's one we all know at least slightly from school: pi, the ratio of a circle's circumference to its diameter. It begins 3.14159 . . . but the numbers carry on for ever. It is infinite and irrational, never ending and never repeating, and many fall into its orbit. To some the attraction may be spiritual, to others the pull may be more like the 'because it's there' reasoning of mountaineers. Memory athletes – so called because of their intensive training – in particular are drawn to the endlessness of pi.

Akira Haraguchi, of Kisarazu, near Tokyo, recited pi to 100,000 digits in 2006 at the age of sixty, a feat that lasted more than sixteen hours. To him pi represents a religious quest for meaning. 'Reciting pi's digits has the same meaning as chanting the Buddhist mantra and meditating,' he says. 'Everything that circles around carries the spirit of the Buddha. I think pi is the ultimate example of that.' He is the world champion, although Guinness World Records has not recognised his recitation.

The official Guinness record holder is 23-year-old Rajveer Meena from Sawai Madhopur district in Rajasthan, India. On 21 March 2015, at Vellore Institute of Technology in Tamil Nadu, Meena recited pi to 70,000 decimal places. He was blindfolded. The feat took him 9 hours 7 minutes. One of the factors motivating him, he told me, was his upbringing. He wanted to show that despite his humble background, he could win the world's toughest memory challenge.

These memory wizards have different motivations, and use different techniques, but they all essentially convert the numbers into a story. When they recite the numbers, they are telling the story in their head and translating it back into digits. Haraguchi uses a system based on the Japanese kana alphabet. The first fifty digits of his system read (translated into English): 'Well, I, that fragile being who left my hometown to find a peace of mind, is going to die in the dark corners; it's easy to die, but I stay positive.' One hopes that in the rest of the 100,000 digits the story picks up a bit.

Meena assigns groups of numbers to words, like Funes in the Borges story. When I chatted with him, he gave me an example. 'I leave my house and meet Roger Federer, go to the park, grab a pair of jeans, get a cab for $50 to the office, where I earn $100.' This translates into the number 749099950100: I (74) leave my house and meet Roger Federer (90), go to the park, grab a pair of jeans (999), get a cab for $50 to the office, where I earn $100.'

Memorising the story for a sequence of 70,000 digits took him over six years. As well as a way to show he was the best in the world, 'it was a good way to increase patience and confidence,' Meena says, deadpan. Those qualities were tested in the seven-month wait he endured before his attempt was officially verified: 'When I got an email from Guinness World Records saying my record claim was successful, that night I was unable to sleep. Many times I checked that mail.'

Wondering if his memory was like that of Funes, I asked him if he recalled everything that had happened to him every day. Meena said no. He has a good memory for faces and incidents but he doesn't inadvertently recall everything he was wearing or eating over every day of his life. Later, however, we will meet someone like this.

If I'm going to understand the people who memorise pi, I feel I ought to get a bit closer to it, so I spend some time scrolling through a webpage listing pi to one billion digits.[2] I've been scrolling down steadily for some time, and the scroll marker in the margin is still only about 5 per cent of the way down the page. I feel my mind could unravel if I do this for too long. The tumbling numbers recall *The Matrix*, but obviously I can't see anything in them, because there *isn't* anything in them. Pi is infinite, and so far has been calculated to 22 trillion digits. No one has (yet?) posted this number online. I go back to the beginning of the sequence, 3.14159 . . . and cut out the first 22,514 digits. It's a mere snippet (though of course even a trillion digits from an infinite number is still a mere snippet), and look through the numbers at greater leisure.

I'm reminded of another Jorge Luis Borges story, 'The Library of Babel', which describes a library containing a monstrous, as good as infinite, number of books, a vast space holding every possible book that can be made, with every possible combination of letters. Overwhelmingly the books are total nonsense, but once in a while a word or even a coherent sentence appears. People in the library spend lifetimes searching for books that mean something. Looking through my snippet of pi I glimpse odd islands of numbers: clustered repetitions of nine, or short strings of zeros and ones that look like binary, or regions where spiky sevens seem to make a snakes-and-ladders zigzag slide through the digits. But that's all. I know there's no meaning here, and I don't really understand those who sink into it. Imagine trying to memorise this number.

To Daniel Tammet the numbers do have real meaning. Numbers to him have an aura. They have colour, texture, shape and even, weirdly, emotion. The number four, for example, to Tammet is blue, but it's also a timid number, one he feels close to because of his own shyness. It's his nickname for himself. Numbers on their own can glow and wink and perhaps even snarl, but strings of numbers form sentences of emotions and feelings.

Tammet, thirty-eight, is a best-selling author and translator. British-born, he lives in Paris, where I meet him on a sweltering June day in a cool, green restaurant in Saint-Germain-des-Prés. I'm doubly testing my memory before I meet him, as I used to know this area really well, so I am wandering the old streets in the heat looking for familiar places (my favourite oyster shop, Huîtrerie Régis, is nearby). I'm also dusting off my French, and finding, happily, that some is still there. I remember fondly a long lunch where I consumed twenty-one oysters at Régis's shop.

Tammet is a polyglot, speaking some ten languages, and also has synaesthesia, a neurological condition which for him means that when he sees words or numbers he also sees colours. 'Three is green, five is yellow and nine is blue – very blue, a different blue to four,' he tells me. 'Tammet' is orange. Rowan is red, Hooper is white. (It's a pleasing combination, I am relieved to hear.) Tammet also has autistic savant syndrome, and his IQ is 150 or 180, 'depending on the scale' they use.

It strikes me as odd that the same person's IQ test results can vary by such a huge amount, but I let it go – I'm more startled by his insistence, like John Nunn, that IQ measures only how good you are at doing an IQ test, and that it 'reduces intelligence to a number'. It's meaningless, Tammet says. We're back with another billionaire who says money isn't everything.

But I'm not here to talk to him about IQ. I want to hear how – and why – he set a European record for the recitation of pi. It took him just over five hours. He recounted it to 22,514 places.

Tammet is the oldest of nine children. He says that the necessity of having to interact with so many siblings as a child tempered the antisocial effects of his autism to some extent. And his autism is relatively mild. He makes as much eye contact as a non-autistic person, but says he has to work at remembering to do so. He had problems making friends and communicating as a child, and his first language was numbers. At school he felt attracted to pi, but was overwhelmed by the idea of an infinite number. In his twenties, still feeling its pull, he printed out twenty pages of pi, one thousand digits per page, and dived in. He basically communed with the number.

'I would look at the numbers and find emotions and shapes — it's like a poem for me, like Baudelaire in French or Shakespeare in English. Pi is like a poem written in numbers. And the further I went into the numbers the more sense it made. Because the more material I had to make sense of it, to give colour to it.' It reminds me of how Haraguchi speaks of finding Buddhist meaning as he goes deeper into pi.

Tammet constructed an emotion-led poem out of pi, and recited it in public, in Oxford. To him it was very much like giving a public reading of an actual poem, or an actor giving a performance. 'It was in the language of numbers, and people in the audience were very touched because although it wasn't their first language,' he says, 'although it wasn't a language they could understand, I was able to embody the number in my body, in my breath, in the way that I spoke. I used numbers as language and this seemed to make an impact on people.'

If Haraguchi recites pi as Buddhist mantra, and Meena for reasons of village pride and record-breaking bragging rights, for Tammet the motivation was more simple: it was about communicating.

During his pi period, did he dream in pi? 'In the moment just before falling asleep I would see the numbers flashing before my eyes, the shapes, the colours, the emotions and the meanings, the particular meaning of feeling alone or feeling scared,' he says. It makes me wonder if bravery is necessary for something like this. 'There are moments in pi when you feel completely on your own. At the first thousand digits you feel like you're the only person in the universe, and it's a terrifying emotion. And then the story continues, you enter a new stage.'

There is physical exhaustion during recitation. Meena says that during the nine hours of his recital he had diarrhoea and fever, and could hardly get the numbers out. Tammet says he doesn't know how actors in long plays do it, but the courage is for the psychological ups and downs that come of a number soaked in emotions.

'There were people who said afterwards they had tears in their eyes, and I must've had tears in my voice,' Tammet says. 'It was strange because I was counting but also recounting . . . it was something that kept me apart from people that finally helped me to communicate directly.' His account reminds me of what Mark Rylance said he had to do after playing Rooster Byron in *Jerusalem*. The performance was so intense and over-whelming, he'd have to sit curled up in a ball after the show and collect himself – draw himself back. 'I needed courage because there were moments when it was unsettling,' says Tammet, 'but I knew there was something beautiful or encouraging coming along.'

Daniel Bor and his colleagues at the University of Cambridge's department of psychology put Tammet through a variety of tests, including a brain scan, and a psychological measure called the Navon task. This involves showing the subject a series of large 'global' characters – letters or numbers – composed of many smaller copies of another, confounding character, the 'local' character. So the subject might see a large number three, made up of small number sevens, and have to identify the number seven. Tammet was faster than average at finding what the local characters were, and was less distracted by the global character.

His synaesthesia is also of a particular kind.

As Bor's paper puts it: 'This appears to generate structured, highly-chunked content that enhances encoding of digits and aids both recall and calculation.'[3] 'Chunking' is the grouping of smaller elements into more familiar units, and is a technique commonly used by memory athletes. For example, the number 10271962 might be remembered as October 27, 1962. Tammet's autism and synaesthesia seem to have aided him in memorising pi.

'Tammet has a predisposition to focus on the details of what he is trying to learn – for instance, the relations between numbers, or the specific features of foreign words – how they sound, what synaesthetic details they may relate to, which he can use to aid his memory,' says Bor. 'I would speculate that this localist approach enhances his particular form of synaesthesia.'

Tammet, in other words, saw pi in synaesthetic chunks that he associated with particular colours and emotions, and he knitted these chunks together into a story. 'He used a mnemonic method, but one intimately connected to his synaesthesia,' says Bor.

People who train their memories are known as memory athletes. Think of them not as Rocky running up the steps of the Philadelphia Museum of Art, but as hunched over a table straining to remember the order of a deck of cards. That's how they train. You can still play the theme from *Rocky* in your head if you like.

There are competitions at country and regional level, and an annual World Memory Championships, run by the World Memory Sport Council.[4] The 2016 world champion was a 25-year-old American medical student, Alex Mullen. Among other things, Mullen was the first person to memorise in under twenty seconds the order of a deck of shuffled playing cards, and the first to memorise more than 3000 single-digit numbers in one hour.

He, like Daniel Tammet and like all memory athletes, uses methods to encode information into a form that makes it more memorable. Our brains don't learn well if we just try and stuff them with raw information – we need to provide a framework of some kind that the brain can feel at home with. That's because the part of the brain involved in processing short- and long-term memory, the hippocampus, is also involved in processing emotion and navigation. For Tammet, the framework was the emotion-driven story he constructed from the numbers. Once he'd remembered the story, he could translate backwards and get the numbers. So he 'chunked' the digits into groups, then mnemonically linked them into a story, in a similar way to how Haraguchi and Meena also wove a story out of chunks of pi.

You don't have to have savant abilities or be aided by synaes-thesia to do this. You just need to practise. Anyone can do it, as science journalist Joshua Foer demonstrated when he wrote about memory championships and decided to learn the skills for himself. He ended up winning the 2006 US Memory Championship, setting a US record in the 'speed cards' event (he memorised a deck of cards in 1 minute 40 seconds).

I talk to Martin Dresler of the Donders Institute for Brain, Cognition and Behaviour at Radboud University Medical Centre in the Netherlands. He's shown that anyone can use the tech-niques of memory athletes to become masters themselves. First, Dresler put twenty-three of the world's most successful memory athletes through a functional Magnetic Resonance Image (fMRI) brain scanner. These athletes have spent hundreds, even thou-sands, of hours practising their memorising methods. Most use a technique called method-of-loci, also known as the memory palace technique. With this method, you imagine a place you are intimately familiar with, typically your house, and you populate a route through the house with items corresponding to the list of things you need to learn. It's like Meena's example of walking to the park and seeing Roger Federer. The more unusual, startling, even unsettling the images, the more memorable they are. By tracing the route in your head, you can pick up the items along the way, and translate them back into the list.

When Dresler's team checked the results from the fMRI, they found no structural difference in the brains of the memory athletes compared to untrained people. They found only differences in brain activity, and then only when the athletes were resting.[5]

Dresler then put volunteers who were new to memory training through six weeks of instruction on the memory palace technique. After this, they had typically doubled their ability to remember words from a random list, and the activity patterns of their brains had started to converge with that seen in the champion memorisers.

Anyone, it seems, can become a memory superhuman. Our potential memory is vast, but the key is to understand how it evolved, and to play to its strengths. 'There has hardly been an evolutionary pressure for our ancestors to store abstract information,' says Dresler, 'whereas memory for visuospatial information – finding the way home, or to feeding or mating places – is crucial for most animals.'

Understanding the evolutionary context of memory is vital to recognising its vulnerability to misinformation, as we'll see. It seems counterintuitive at first that to remember more information we need to encode data into bulkier shapes – we need to build palaces filled with penguins and space stations and Roger Federer; we need to create more information to remember less – but Dresler explains why this is.

'Our brain is biologically still mainly prepared to encode very concrete visuospatial information, and is much less suited for abstract information. Encoding via an additional step of transforming abstract information into visuospatial representation, and thus the encoding of additional information, is more efficient than encoding abstract information directly.'

We learn by listening to stories, and if we create stories we can learn to remember. All the memory athletes Dresler has worked with say they have no innate skill. Everything they do has been learned. But there is a form of extraordinary memory that you do just seem to be born with.

I'm waiting in front of a huge bronze head of Nelson Mandela. The bust is next to the Royal Festival Hall on London's South Bank. There are a few other people nearby, a woman with red hair and a man in a raincoat, wearing black-rimmed glasses. Other people stop to take photos next to the statue. I'm wearing black jeans, leather boots, a striped long-sleeved cotton shirt and a khaki coat. I'll spare the details of my socks and underwear, though I will note that there's a (non-fatal) rip in the crotch of the jeans that I made when I was climbing through the labyrinth in the Jardin des Plantes in Paris last week. I study Mandela while I'm waiting. He's young in this statue, more the defiant leader than the elder statesman. There is cloud in the sky threatening rain. Now a fair-haired young man approaches. He's taller than me, wearing black jeans, with a pale shirt open over the top of a black T-shirt. He has a small silver hoop earring in his left ear. He holds out his hand and says my name.

If I hadn't written that down the morning after I was there I would soon have forgotten those details. And it happened only yesterday. Ask me what I was doing on Monday two weeks ago, and I'll struggle to remember, let alone recall what I was wearing and what the weather was like. Yet some people remember these sorts of details, and more, for every day, going back years, even decades. They have a condition called highly superior autobiographical memory, or HSAM.

It was first officially described in 2006, when a woman called Jill Price got in touch with James McGaugh at the University of California, Irvine. It turned out Price was able to recall, accurately, huge amounts of frankly unremarkable information like that I listed above – for every day of the last thirty years. 'Starting on February 5th, 1980, I remember everything,' she said. 'That was a Tuesday.' You could throw a date at her – and McGaugh and his team threw plenty – and she would tell you the day of the week and what she did. For example, 3 October 1987? 'That was a Saturday. Hung out at the apartment all weekend, wearing a sling – hurt my elbow.' The scientists were able to compare details against a set of journals that Price had kept over the years, and check with calendars the day of the week she gave. She indeed has an unfailing memory, one quickly dubbed 'total recall' by the media.

Since Price first turned up, more people with HSAM have been verified, including the actress Marilu Henner (she played Elaine in the 1980s sitcom *Taxi*). HSAM is an innate kind of superior memory; it is unlike learned abilities such as the recounting of pi, which use mnemonics. Somehow these HSAM people just do it.

I want to understand more about it. *How* do they do it? What does it feel like, to have a head so crammed with memories? 'Most call it a gift but I call it a burden,' Price has said. 'I run my entire life through my head every day and it drives me crazy!'

Maybe people with HSAM can help tell us what a memory is, and how they are stored. I remember seeing a Harry Potter film in which Dumbledore extracts memories from Harry's head with his wand. It's hard to rid myself of the image of a memory as a wispy, tendrilous structure. What form does a memory have, and how do we access them? I ask Julia Shaw, a psychologist at the London School of Economics, and author of *The Memory Illusion*. 'A memory,' she says, 'is a network of neurons that are physically connected and "hum" together, that is, fire on the same wavelength.'

A network is a physical structure, so a memory does have actual form. It's like a spider's web in the head. A memory is activated by sending out a 'probe' – like casting a fishing line with an enquiry on the end instead of a worm. Say the enquiry is about a beach. It activates related memories. 'The concepts that are strongest in their relationship to that probe are going to be activated automatically,' says Shaw, 'for example, "beach I recently visited in Florida". And when they are recalled, it is at this point we can check the memory to see whether it is the one we are looking for.' Perhaps I don't want to think about that *Florida beach*, so I refine my probe to make it about *black sand beaches*, and I come back with the right memory about a *black sand beach in New Zealand*. I'll bear this fishing line idea in mind.

The taller-than-me fair-haired young man shaking my hand is Aurélien Hayman, and he has HSAM. He is twenty-five. Since around the age of fourteen, he's been able to remember every day of his life. We have a beer at a quiet table inside the Royal Festival Hall. The bottle had a sky-blue label, it was a local pale ale, brewery name of . . . no, it's gone.

With Jill Price, who said her memory is 'nonstop, uncontrollable, and automatic',[6] the ability seemed to come about overnight, but Hayman is hazy about when his kicked in. 'I don't remember when it happened. I always thought it was like a party trick, just something I could do. It's not like I banged my head and it happened.' He saw a documentary saying a handful of people in the world had this ability that they could remember every day of their lives, thought, 'this is me,' and Googled it.

This led, domino-like, to a meeting with memory researcher Giuliana Mazzoni of the University of Hull, scientific studies, media articles, TV appearances and a Channel 4 documentary.

I get in touch with 27-year-old Rebecca Sharrock from Brisbane, Australia, who has an extreme, almost Funes-like form of the phenomenon. 'Whenever I'm reliving a memory it is vividly detailed,' she says. 'All of my emotions attached to the event come back, as well as all of the things that my five senses picked up from the experience.'

Asked to give an example, she says that each year on her birthday she typically flashes back to one of her happiest memories. It is her seventh birthday. 'I can smell jasmine in the air, and my mind is full of pictures of a pink and gold sunrise. There is also an echo of the excitement within me from before I opened my presents.' She goes on to list the presents she got that year, 1996: a princess tiara, a toy pony and a model house. But there are also what she calls intrusive memories – involuntary recollections of moments of physical and emotional pain. She can remember grazing her knee, and with the recollection comes a phantom echo of the pain. 'Though as unpleasant as pain is, reliving negative emotions is so much worse.'

For a long time, having HSAM was gruelling. 'I believed that everybody remembered in the same way I do,' she says. 'I just thought that most people were better at handling their emotional flashbacks than me. This made me feel broken and depressed.' But now she knows – through visiting Craig Stark's lab at the University of California, Irvine – that she is not alone with the condition. 'Now I feel so much happier.'

Sharrock is unusual too in that she has autism, and has memorised the entire series of Harry Potter books. They helped buoy her when she was trying to sleep and memories were flooding in. I ask her to tell me a favourite scene. She says that given what we're talking about, it has to be when Harry persuades Professor Slughorn to hand over his true memory of what he told the teenage Voldemort. 'That section means a lot to me because Slughorn accurately describes how I myself feel about shameful memories.'

Stark says Sharrock is a good illustration that HSAM comes in strong and weak forms: 'There is also a range of autobiographical ability. Some perform better on the tests than others, but all are far, far beyond the typical individual.'

The way Hayman describes *his* memory, it isn't a burden. 'It's often expressed as "gift or curse?" but for me it's a very small facet of me, of who I am. When people say "how does it affect your day-to-day?", well, it doesn't really.' Hayman's HSAM is entirely under his control. 'I don't wake up in the morning and go, "oh it's June 5th" and think of all the other June the fifths. If someone asks me a date going back, sometimes I can't remember it. There are blank patches.'

This seems like a good time to throw him a random date. Let's say, 1 May 2005. There's a long pause. He's staring out the window, obviously looking into the past. I imagine him rifling through endless filing cabinets. 'That's not something I can immediately recall,' he says.

I'm forgetting that in 2005 he was only thirteen. I don't know what I expected – computer-like recall of a date and an immediate report of the things that happened on that day? – but he continues to reach towards the details of the date I have given just like we all do when trying to remember something. Then it clicks, and he snaps his fingers. 'I've got it, completely, solidly,' he says. Suddenly the picture has resolved. 'It was a Sunday.' (I check this later and he's correct.) 'Me, my dad and my mum went out to the Chilterns for the day and had a pub lunch. I can remember loads of stuff about that day – clothes and weather – but it's probably mundane for you.'

I ask him to describe the moment just then when he was trying to remember the date.

'If the memory is not immediately obvious there are sort of checkpoints in my mind, and I was thinking, "my birthday is April 27 and I knew it was the bank holiday around then" . . . you sort of sift around in your head and find it.'

This sounds like what Julia Shaw said about probes going out and seeking related memories. When I think about memories, I find they are grouped by my physical location at the time, say my apartment in Tokyo, or my flat in Dublin. I find it much harder to order my memories by time rather than space, but that's apparently what people with HSAM can do. 'For people with HSAM dates are the optimal cue, but there seems to be a sort of chaining effect, that is when they retrieve something thanks to the date, they are able also to use some of the elements that they have retrieved as cues for other memories,' says Mazzoni. If enough time were given to them, she explains, they might retrieve much more, even if not every instance of their life.

'People think that this is all about having limitless memory, a really expansive capacity to remember all this stuff,' Hayman says. He doesn't believe, though, that his memory is more extensive than other people's; it simply seems to be arranged in such a way that he can pick things from it at the right point. 'There's some kind of accessibility which is really fine tuned for me.' Given enough prompting, he reckons, I would be able to recall what happened on a given date, too. Maybe, but given enough prompting I might also just *imagine* I could remember the event, and this can have devastating consequences, as we'll see. 'In the right conditions many people should be able to retrieve more,' says Mazzoni. It's why she finds HSAM so fascinating: it gives us a glimpse of the potential capacity of long-term memory. Based on both HSAM and other phenomena, she believes, we probably have mental representations of many more experiences than those we are able to retrieve at a given moment.

Okay, so I see how Hayman can grasp his way to the memory. But how does he remember what day the memory relates to? I struggle with this at first because when I think about myself I'm aware that I don't know what day it is every day. So how can I remember years back? Hayman shrugs, and agrees – he doesn't know how that works. But thinking about it, even if I go through a day without explicitly saying the name of the day of the week (as I write this a memory from *The Godfather* pops unbidden into my head, Michael Corleone's Sicilian wife saying 'Monday, Tuesday, *Thursday*, Wednesday . . . '), even if I don't think I know what day it is, of course I do really, deep down. It's just taken so much for granted, like the sun and the rain, like breathing, that we don't think we take note of it.

I throw another date at Hayman – 9 July 2012.

'Let me think about this.' He goes into recall mode, staring out the window, and I'm picturing tendrils of thought snaking out and gripping hold of markers poking out of the mist.

'It's a Monday.' Yes, I later confirm this. 'I remember . . . that was the day I was being filmed for the documentary.' He laughs that I've alighted randomly on that day. 'I remember what I was wearing but that's because it was on the documentary. It was Wimbledon the day before, my hayfever was really bad at that time. I can tell you the weather if you want.' It *was* the Wimbledon final the day before: Roger Federer beat Andy Murray.

I wonder how far the memories go. This is ridiculous, I say, but could you remember how cloudy it was, or even the shape of the clouds in the sky? Hayman gamely smiles. He can't. It's not photographic in that way, he says. 'Sometimes it's just a sense of what it was like, a mood.'

Indeed, I should know better. I was getting carried away with thoughts of Ireneo Funes. Even Rebecca Sharrock doesn't remember that sort of detail. There is no such thing as 'photographic memory'. The closest thing to it is eidetic memory, which is the ability, seen only in 5 per cent of children and never in adults, to view a memory of something as if it was a photograph in front of them. Say you show such a child a room for a minute or two, then blindfold them completely. For just a few minutes they will be able to view their memory of the room as if it was still in front of them.

Some people with HSAM seem to have better-than-average working memories. What's Hayman's like? 'It's really bad. I am actually quite forgetful. I'll forget to do something at work. It's a completely different thing.'

HSAM is weird in that people with the condition, or ability, can remember only stuff that they personally experienced, and they get the recall ability only some months after the experience.[7] It's not like they've immediately encoded their memories in some crystal protective shell. It's more that the route back to the memory has to have time to bake in. The experiments that showed the existence of this delayed-baking phenomenon were carried out in Craig Stark's lab. 'One of the take-homes from the study is that many aspects of their memory are entirely typical – anything outside of autobiographical memory – and that the fundamental mechanisms involved appear typical as well,' Stark says. 'It's not that people with HSAM are wholly different from us and using some system entirely unknown to the field, but they're clearly performing in ways that are orders of magnitude beyond the rest of us – in this limited domain.'

Time for another date for Hayman. What about 12 March 2009? He struggles with this one, then says he can do 14 March. 'But maybe this is a memory I shouldn't share.' Oh, go on. 'It was a Saturday. It was the first time I ever got drunk. We were on the beach, Penarth Beach.' What were you drinking? 'Straight vodka.'

Hayman's memories are not all of the same strength, just like those of us without HSAM. Some memories are quite diluted, some are really sharp. I imagine he snapped towards the 14 March memory of getting drunk because it was much sharper (at least until he got drunk) than the 12 March memory. And all 'firsts' are memorable. But generally he doesn't know why memories vary in this way. Perhaps it's to do with how much they are pondered afterwards.

Hayman shares another characteristic with many people who have HSAM – not that there are lots of them; only some sixty people in the world have been identified with the condition. 'I have a very active imagination. I'm naturally a bit of a day-dreamer.' It doesn't seem to be the guided daydreaming that Hilary Mantel spoke of. This is much more . . . unmoored. 'Sometimes I couldn't even tell you what I'm thinking about,' he says. 'My parents might say I'm away with the fairies.'

This could be key to their ability, says Lawrence Patihis – formerly another of the UC Irvine group, but now at the University of Southern Mississippi at Hattiesburg. In a paper on HSAM personalities, Patihis wrote: 'The personal events in which HSAM individuals become absorbed in and fantasise about later will likely lead to accurate memory.'[8]

Hayman repeats that it's a misunderstanding to think of HSAM as a burden, at least in the form that he has it. 'It's not like having your entire life history that keeps playing in a reel in your mind, it's not like that at all. It's just, if prompted, you can recall things. It's a talent of accessing memory rather than just possessing loads of it.'

His feeling is that he has more memories there somewhere, only even he can't access them. He thinks the rest of us do too. Perhaps everything we experience is encoded somewhere, but locked away, or neglected and so faint that we can no longer access it. 'Could I,' wondered Dorthe Berntsen of the Department of Psychology and Behavioural Sciences at Aarhus University in Denmark, 'as a non-HSAM person, have memory from each day in my life stored, but I just can't get to it?'[9]

It's a compelling and widely held view, but it's a widely held myth. When I mention it to Patihis he slaps it down, hard. 'Absolutely not. That is a long-debunked idea going back to the neuroscientist Wilder Penfield.' Penfield, a Canadian neurosurgeon, was the first to map and deduce the function of various parts of the brain, and he pioneered surgical treatments for epilepsy. 'Most things we experience are not encoded,' says Patihis, 'and even those things we do encode fade over time.'

Dutch neuroscientist Martin Dresler points out that it's hard to think how you would even test this idea. And in any case, from an evolutionary point of view it doesn't make much sense to equip the brain with a vast capacity to encode everything, then conceal it from conscious access. 'I think it would make much more sense to encode only experiences fully and permanently that were somehow tagged,' he says, 'for example by strong emotions, as very important, and to keep only the gist of more mundane experiences, getting rid of all redundant boring details.'

The question, according to Hayman, is how he can access those bits that the rest of us can't. The information he retrieves isn't really significant. But you can't blame the brain for not knowing what's significant and what's not. A pub lunch with Mum and Dad in the Chilterns, there's at least *potential* for significance in that. But then there's potential in everything. It's only if we tattoo the memories in, for weddings and special occasions, or use mnemonics, or practise over and over, like Hafiz devotees learning the Koran, like actors learning their lines, that they become more deeply engraved.

Psychologists class memories not by emotion, as we tend to (happy memories, sad memories), but more broadly, on whether the memory is about general knowledge, or about personal experience. The former are known as semantic memories, the latter episodic. It is these personal memories that are highly detailed in HSAM people.

Experiments carried out by Patihis and his colleagues have shown that HSAM people may have amazing chronological recall, but the detail of the memories is only average.[10] They are no better than non-HSAM people at remembering, say, the words in a list presented in a laboratory setting. This is consistent with Hayman's own feelings about his memory. 'I can remember things in a patchy way and I can remember things around it. If you said a month in a particular year I could remember the kinds of things I was listening to, the friends I had, what was on the radio. If you threw a date at me it might unlock a memory for the first time since it happened. Which is bizarre.'

Talking with Hayman has been illuminating, not least because it has demystified what HSAM people do. Quite probably they use the same memory mechanisms as the rest of us – think of the enquiry cast out on the fishing line – and there is evidence of this from a false-memory experiment carried out by Patihis and his colleagues. In these psychological tests people are shown scenarios and then tested on them in such a way that they are seeded with potentially false information. All of us are susceptible to blithely incorporating the fake news into our memory. 'The memory distortion tasks used in that study really do tap into the variety of ways we reconstruct memory,' says Patihis, 'suggesting that storage and retrieval of memories in HSAM people is similar to the rest of us.'

He thinks it is not so much new memory mechanisms that explain HSAM, but personality differences such as obsessiveness, a vivid imagination and the tendency to become absorbed.

People with highly superior autobiographical memory, then, are not *so* highly superior after all. Certainly, we don't know how they fish out so many memories from thousands of days stretching into the past. But just like the rest of us, they are susceptible to contamination – to the adoption of and belief in false memories. Indeed, the very thing that might be responsible for their superior memories – their absorption and tendency to daydream – could be what leaves them prone to misinformation. Deep absorption in misleading information can produce errors in memory, says Patihis.

HSAM may be characterised as total recall, but if you remember the movie of the same name there's a detail that rather spoils the comparison: Arnie's memories have been faked. He misremembers vast swathes of his life. The incorporation of false memories into our brains can have tragic consequences, but its widespread nature is poorly recognised. Time then to look more deeply.

Ledell Lee was executed in Arkansas in April 2017 for the murder of his neighbour, Debra Reese. However, according to the Innocence Project, a legal organisation that campaigns to exonerate wrongly convicted people, there were numerous problems with his conviction. For example, many unknown fingerprints were found at the crime scene, but none of them were Lee's. DNA testing of a speck of blood on Lee's shoes was not carried out, nor on hairs found at the scene which the prosecution argued came from Lee. The prosecution case also relied heavily on eyewitness testimony.

Three people identified Lee as a man they had seen in the area and leaving Reese's house. Yet according to a report by the Innocence Project, of 349 exonerations of convicted prisoners by DNA evidence in the US to date, 71 per cent had rested at least partly on eyewitness (mis)identification. The US Supreme Court did not order a DNA test for Lee, and he was executed by lethal injection.

Psychologists and lawyers have long known that eyewitness identification is unreliable. There have been thousands of scientific papers demonstrating problems, not least something called 'own race bias'. This is the tendency to be worse at identifying people from ethnic backgrounds different to our own.

But eyewitness identification continues to be a powerful strand of evidence; the science, outrageously, hasn't filtered down to the courts of law. In one infamous case in 1984, a young college student was attacked and raped in her apartment in Burlington, North Carolina. The victim, 22-year-old Jennifer Thompson-Cannino, who is white, said she'd specifically tried to memorise her assailant's face as he was assaulting her, so that she would have a chance of identifying him if she survived. Local restaurant worker Ronald Cotton, who is black, was at home the night of the attack, but fluffed his alibi and Thompson-Cannino picked him from a selection of police photos. When Cotton was brought in for a live line-up, he was picked out again by Thompson-Cannino, who said she was 100 per cent certain he was her attacker. It's hard to argue with an assertion of total certainty, and we are prone, both as jurors and simply as people, to place greater weight on something told to us by someone else than on dry scientific evidence. Her testimony at trial helped sentence Cotton to life plus fifty-four years. The story has a happier ending than Lee's, however: just over ten years later DNA evidence exonerated Cotton (and implicated another man, who had confessed to the crime in jail). Thompson-Cannino and Cotton became friends, have written a book together (it has the great title *Picking Cotton*), and regularly speak on the need to reform the law relating to eyewitness identification.

'Even when witnesses say they are 100 per cent certain of something, studies show that the testimony is only slightly more accurate than people who say they are unsure,' says Jakke Tamminen, a psychologist at Royal Holloway, University of London. There is a positive correlation between confidence and accuracy under pristine conditions, but not in the messy real world. And no one, least of all prosecuting lawyers, takes this into account.

At Tamminen's lab I take part in an experiment to test memory reliability. He has me witness a couple of crimes, staged for the camera for purposes of memory research. In the first crime, two men steal a computer monitor from a library; in the second, a woman in snowy Cambridge, Massachusetts, has her purse stolen by a man on the street. There are lots of complicating factors in each scenario – other people interacting with the main characters, and several different locations in each. After I've watched the crimes, I read a list of statements summarising the plot of each scenario, and the actions of the people involved. Tomorrow I will be examined on the facts of the crimes. The idea, says Tamminen, is to test the accuracy of eyewitness testimony.

It's been shown experimentally many times that even if we have vivid memories which we rate as highly reliable and highly likely to be true, this does not mean that they are necessarily accurate. The morning after the space shuttle *Challenger* exploded in 1986, Ulric Neisser, a cognitive psychologist at Emory University, had students fill out a questionnaire about the disaster. Three years later the students filled out the same questionnaire and Neisser compared the answers: they were completely different. Similar comparisons, conducted following the 9/11 attacks on the World Trade Center in New York, found complete mismatches between what witnesses were sure they remembered and what really happened. Our minds quite literally play tricks on us.

The morning after I've witnessed the 'crimes' in Tamminen's lab, I am questioned about what I saw, and asked to rate my confidence in each answer. For example, the woman who had her purse stolen. When I read the summary statements the day before, I was told that the man bumped into her from the front. I thought this was misinformation, because I remember thinking it looked unnatural when I watched the crime that the man had bumped into her from behind. The next day, I am again told that the man who stole from the woman bumped into her from the front, and am asked to agree or disagree, and rate my certainty. I disagree, and say I am fairly certain (4/5 on the scale). But then I am told, 'the man put the purse into his jacket pocket', and I can't remember if he did put it there, or if it went in his trouser pocket. The same with other details — was a woman in the library wearing glasses? Tamminen's experiment is designed to test how certainty about memories changes depending on whether they are tested soon after witnessing an event, or with a period of sleep in between.

We know that when even a long-term memory is retrieved, it becomes liable to rejigging. The demonstration of this caused quite a stir when it was published in *Nature* back in 2000, as researchers thought long-term memories were locked down.[11] And work by Jason Chan, at Iowa State University in Ames, shows we are more susceptible to being contaminated by false information when recounting an event straight after it occurs. The act of remembering itself makes a memory more unstable, and this is when misinformation might sneak its way in.[12]

We all seem to be prone to believing things that aren't true. To believing, faithfully but incorrectly, that things we didn't witness did in fact happen in front of our eyes. Maryanne Garry, a psychologist at the University of Waikato in Hamilton, New Zealand, has studied the acquisition of false memories for many years. 'My own hunch,' she says, 'is that very few, if any, people would be immune to this effect given the right package of information, method, and circumstances.'

Jason Chan agrees. 'Some people are a little more resistant to false memories than others, and frontal lobe functioning – which is related to source monitoring ability and working memory capacity – is typically thought of as an important factor.' However, resistance is one thing; immunity is quite another. 'We are not aware of anyone who is not susceptible to false remembering at all.'

So are there any superhuman resistors? If anyone could be trained, or selected to be resistant to the tendency of our brains to soak up false memories like blotting paper, you'd think it would be the military special forces, whose training is particularly intense. Many countries train their elite soldiers to resist interrogation.

Charles Morgan III is a psychiatrist at Yale School of Medicine who has worked extensively with the US and Canadian military to advise on eyewitness memory and psychological performance under conditions of high stress, and to aid in selection for special forces. Subjects are typically serving military personnel undergoing special forces training. In the studies, soldiers take part in an intense role-play exercise modelled on prisoner-of-war conditions, being deprived of food and sleep for forty-eight hours, and then being 'robustly' interrogated. A day after their 'release' from the POW camp, the soldiers are asked to identify their interrogator in a line-up. Several of Morgan's studies have found that many soldiers are unable to, some even mistaking the interrogator's gender. 'After they are "interrogated" in what sounds like a brutally realistic way, large numbers of them just totally suck at recounting what happened, and at identifying their interrogator,' says Maryanne Garry.

For special forces soldiers it may not be a good thing to be susceptible to memory distortion, but most of us will never be captured and placed under intense stress. We have not evolved that way, and indeed our brains work differently. We are social apes; we learn from a range of different people, at different times and under different conditions, as well as from various cues in the environment. Our brains have to be malleable. 'I think it's pretty straightforward,' says Garry. 'We have evolved to learn from multiple sources. We have not evolved to detect errors in the way we often encounter them in real life . . . from sources that otherwise seem legitimate, or trustworthy or not worth the brainpower to scrutinise.'

Nevertheless, there are suggestions that some people are more resistant, at least.

Bi Zhu, of the Institute for Brain Research at Beijing Normal University, China, has worked with UC Irvine's Elizabeth Loftus, a legend in the field of eyewitness memory, on testing whether some people have more reliable memories. Zhu ran experiments similar to the one I took part in at Royal Holloway. She had 205 Chinese college students watch staged 'crimes', and then had them answer questions on the crimes, while seeding them with misinformation. The first set of answers were given an hour after viewing the crimes, but the second test came a year and a half later. MRI scans were made to measure the volume of key structures in the brain. Students with larger hippocampi remembered more questions correctly, and were less susceptible to the false information.[13] What we don't know is how much genetics influenced the hippocampus size of those students.

The realisation that memory is unreliable has profound implications, quite literally, for who we think we are. Your life is like a Russian doll, with other individuals nested inside you. If the you from one stage of your life could speak to the you of now, they wouldn't agree on what each had experienced. At first this concept troubled me, but now I feel quite grateful to evolution for providing us with a method of tweaking our personal histories. I heard a 63-year-old man the other day say he was finally happy with the person he was. It's thanks to our malleable memories that most of the time we are able to grow into people we like.

Felipe De Brigard, who, tellingly, is from the department of philosophy *and* the Center for Cognitive Neuroscience at Duke University in North Carolina, has a startling idea.[14] Memory isn't just for remembering, he says. Misremembering is so common it shouldn't be seen all the time as a malfunction. In his view, many cases help us construct scenarios of past events that might have happened, so as better to simulate possible events in the future. An unreliable memory may also destabilise your personality. You may think that your personality is something unchangeably intrinsic to you, but a study in 2016 that measured personality traits over a sixty-year period showed they can profoundly alter over a lifetime.[15]

Examining superhuman traits expands our understanding of the diversity of the human species. Unexpectedly, this exploration of memory has shown that we are ourselves more diverse than we think. It doesn't seem to make sense to say an individual can be diverse. But the truth is that each one of us contains, within our self, different people.

3

LANGUAGE

Language comes into being, like consciousness, from the basic need, from the scantiest intercourse with others.

Karl Marx, *The German Ideology*

'Mario, what do you get when you cross an insomniac, an unwilling agnostic and a dyslexic?' 'I give.' 'You get someone who stays up all night torturing himself mentally over the question of whether or not there's a dog.'

David Foster Wallace, *Infinite Jest* (2011)

There's a large green parrot in the room next door. I'm in an apartment in Thessaloniki, in north-eastern Greece. I am concerned that the parrot might squawk all through the night and keep me up, so I go to say hello to him. 'γεια σου, γεια σου,' I say, through the bars of his cage. 'Yassou yassou,' exhausting my entire vocabulary of Greek in one go. The parrot eyes me and I actually see his pupil dilate. Then I experience the blip of serotonin-pleasure that bilinguals must get all the time when they are understood by a native: the parrot says hi back. 'Yah, yah!'

The encounter is a fitting one, or so I tell myself. I'm in Greece to attend a conference of polyglots: people who can speak multiple languages. One of the key reasons to write this book was to meet people who are the best at things we care about. To meet them and learn from them. And unless you are extraordinarily self-assured, you care about being understood. We are a social species. We thrive on communication. The better we communicate, the more we talk, the happier we'll be and the more friends we'll have. Perhaps it's not too much to say that with better communication our lives will be more prosperous and the world itself will be safer. It sounds like advertising patter, but I tend to believe it.

I'm far from being a polyglot myself. I do speak Japanese, which is unusual for a Brit, but then I did live in Japan for eight years. During that time it became very clear to me that the more I learned of the language, the more I enjoyed my life there and appreciated the culture, and got on with the people. I used to be pretty good at French, mainly because I liked it at school, and because I spent three summers there for the field-work component of my PhD. It's still there somewhere, in my head. I also did German at school but didn't really get on with it at the time (though I like it now).

So I've come here to learn from the people who love to talk so much that they haven't restricted themselves to a single language. Some of them speak dozens. You know when you're the embarrassing tourist on holiday unable to speak the language? That's me at this conference. I'm the least lingual person here by far.

In this chapter we'll meet people who can slip effortlessly between languages. We'll examine the evidence for the advantages of being a polyglot, for the best way to learn new languages, and consider what genetic differences polyglots might enjoy over the rest of us. We'll also see how exposure to multiple languages was the norm during our evolution and for most of our history, and is still the norm rather than the exception today. We'll find out what this means for our brains — for the rate of our development as children and for our inevitable decay as adults – and even for our system of morality, and for the evolution of language itself. Finally, we'll ponder what it means for the future, given that world languages are going extinct at the extraordinary rate of one every three months.[1]

We'll start with Alexander Arguelles, a tall, middle-aged American of stern, stately demeanour. When I meet him in Greece he is wearing a burgundy-coloured shirt that my mind's eye wants to extend into a cloak: he has an imperial air about him, an aura of knowledge. He also has about him a small cluster of fans. People lower their voices in awe when he is near; some braver ones ask for a selfie. He is, someone tells me, one of the most multilingual people in the world. Arguelles has studied sixty or seventy languages and retains a high level of comprehension in at least fifty. He is no mere polyglot. For him the word is too small. He is a hyperpolyglot. (The term was coined by a British polyglot, Richard Hudson, in 2008.) You generally earn the hyperpolyglot distinction if you know more than eleven languages, although the International Association of HyperPolyglots offers membership if you're fluent in more than six.[2] In this world Arguelles is a legend; he is the grandfather of the polyglot movement. It's no wonder that he attracts such attention.

I wait for my turn to speak. Everyone at the meeting is wearing cards around their necks with flags on them, and the invitation: 'You may speak to me in these languages . . .' Our conference pack contains a sheet of stickers of various flags, and the idea is you adorn your name cards with the flags representing the languages you know. I have the British and the Japanese flags. Arguelles' card is completely covered, on both sides, with stickers. Many of the flags I don't even recognise. He is chatting in Korean with someone.

All around me people are chatting in different languages, and breaking off and picking up in another tongue. How do they do it? Someone tells me it's like a roulette wheel rotating round and clicking into the next language. But with some it's as if an invisible shutter flits over them, drawing across them the curtain of an entire new culture, complete with associated body language and inflexion and gesture and facial expression, and knowledge of social mores. Arguelles is, by his own admission, a lone studier. I find at the conference that there are two types of polyglot: those who plunge themselves into foreign society and pick up the language that way, or those who prefer to shut themselves away with books and study their way to fluency. It's somehow fitting that we have our chat in the near pitch dark (we can't find a light switch) of a backstage corridor.

His motivation, he says, has been to learn languages so he can read the literature of certain cultures in their original language. It's one reason he's studied several dead languages, such as Latin and Old Norse. As an undergraduate and graduate student in the US, he learned Old French and German, and also Latin, Greek and Sanskrit. Moving to Berlin for a post-doctorate he learned other Teutonic languages such as Swedish and Dutch. But needing more of a challenge, he took a position at Handong University in South Korea. He calls this his 'monastic' period, when he could realise his dream of becoming a polyglot. He shut himself away in an isolated house and studied dozens or even scores of languages every day, sometimes for eighteen hours a day. He success has been remarkable. He knows Arabic and Afrikaans, Swahili and Hindi, Irish Gaelic and Persian, Russian and Icelandic . . . well, I won't go through them all.

'I was destined to be a polyglot,' he says. His father, a university librarian, was one, and Arguelles would see his dad studying each morning. But his father didn't encourage him in languages, and although the family lived in various countries around the world, Arguelles grew up a monoglot. There is the merest hint of sadness in the way he tells the story. It makes me wonder: does he think he became a polyglot because he wanted to be like his dad? Or does he feel there was a genetic effect too? In other words, there's a chicken-and-egg question to answer here: did he happen to have the right kind of brain to excel at language learning, or does he put his success and desire to learn more down to the influence of his father's vocation?

'If there is a genetic link it comes from my maternal grandmother,' he tells me. His grandmother, a German immigrant's daughter in the Midwestern US, grew up bilingual in English and German. Falling in love with Spanish, she taught herself the language as a young girl to such a level that she got a scholarship to study in Mexico. She also learned Portuguese, becoming a professional translator and interpreter of four languages.

Of course, a familial influence can operate both genetically and environmentally. There's no 'polyglot gene', but he may have inherited something helpful from his grandmother. And what about his own children? Arguelles first answers by saying he *now* has a great 'polyglot relationship' with his own father. I take this to mean he converses with his dad in various languages. He himself is not the same sort of parent as his father was to him. 'I speak French all the time with my sons, and I'm teaching them Latin, German, Spanish, Russian. They're having a different experience.' Well, this is a chapter about language, not about men's relationships with their fathers. As always, there's both nature and nurture going on.

Arguelles compares learning languages to physical training. 'You can approach polyglottery as a sport, as athleticism, as mental exercise. As rules and sets of things you follow in order to do them. Playing games is fun, isn't it? So it's fun, and it's fulfilling, and it gives you happiness. And there's many things out there that can make you happy but take it from me, there's nothing better, the most fun thing in the world is autodidactic learning.'

Hyperpolyglots can seem otherworldly. They look like anyone else, but have in their power a form of shape-shifting: they have the ability to deploy this superpower of communication and blend in to a different culture. It's no wonder intelligence agencies like to recruit linguists.

The brains of polyglots and hyperpolyglots are different from those of monoglots. We know this from a number of sources. First, from dissection of the brain of Emil Krebs, a German diplomat who lived from 1867 to 1930. A legend in the polyglot world, he could converse in some sixty-five languages, from Arabic to Mandarin to Turkish to Hebrew to Greek to Japanese. He had the extraordinary ability to pick up languages with outrageous speed: Armenian from zero to a good level in two weeks, for example. When he died his family was persuaded to donate his brain to the collection at the Kaiser-Wilhelm Institute for Brain Research. In 2002, when the brain was examined by neuroscientists, they found architectural differences in Broca's area – a region in the frontal lobe known to be involved in language function – compared to the brains of normal people.[3] The hemispheres of the brain were more asymmetric than usual too. So his brain was different – as well it might be. *He* was different.

Learning a new language exercises the brain in a similar way to how jogging exercises the body. It builds up those parts of the brain that are being used, and keeps them flexible. This presents us with another chicken-and-egg conundrum. Do polyglots develop different brains because they exercise them a lot, or were their brains different to start with?

Johan Mårtensson at the department of psychology at Lund University in Sweden has examined this question in a couple of clever studies. Until recently, Swedes faced compulsory conscription into the military, and language experts could apply to the Swedish Armed Forces Interpreter Academy. Selection criteria were very tough: from between 500 and 3000 applicants, only thirty would be accepted, says Mårtensson. He and his colleagues looked at the brains of these select conscripts, and measured the size of the hippocampus and the thickness of the cortex in the interpreters before and after three months of intensive language training. For control subjects, they used students studying medicine at Umeå University – the idea being that both the experiment and the control groups are comprised of young people studying hard, but only one group is learning a new language.

The volume of the hippocampus, it was found, grew during the language training, and grew most in candidates who achieved the highest levels of ability. The hippocampus is a paired structure, supposedly resembling a seahorse (from which it gets its Latin name), hidden under the cerebral cortex, the part of the brain concerned with the higher functions relating to consciousness. It is one of the first parts of the brain to decay when Alzheimer's takes hold. I saw this for myself once while observing a brain dissection at the Division of Brain Sciences, Imperial College London. The brain, already removed, is carefully sliced for examination, and the slices are laid out like ham at a deli counter. The pathologist pointed out how the hippocampus of the dead man was atrophied, shrivelled.

Mårtensson's team didn't have to wait for their subjects to die before they could measure their brains. They used fMRI brain scans. The scans showed that the subjects who had reached the highest proficiency level had greater malleability in the right hippocampus, and in a part of the brain called the left superior temporal gyrus, a lobe of the brain located, as you look at the side of the head, above the ear.[4]

In another before-and-after study, Mårtensson looked at brain changes in Swedes learning Italian over a ten-week course. This time the subjects weren't language experts before the trial, they were normal Swedes recruited from adverts. Again, the team found that the right hippocampus changed – in particular in its structure of grey matter, the bodies of the brain's neurons – compared to the control group who didn't learn a new language.[5]

So learning new languages does rewire the brain. How does this relate to the studies showing that Alzheimer's is delayed in bilingual or multilingual people? It seems that the fattened hippocampus is able to hold out longer against the ravaging effects of the disease. Irritatingly for those who are 'only' bilingual, however, the evidence is not clear that two languages is enough to provide the protection.

Morris Freeman of the University of Toronto in Canada and colleagues reviewed the literature on the protective effect. Some studies, they found, showed that learning two or more languages delayed Alzheimer's by up to five years, but others showed you needed to study four languages to get the effect.[6] The protection is thought to derive from something called cognitive reserve. This refers to brain changes – such as those we've seen in the hippocampus in Mårtensson's studies – that can help compensate for the damage that comes about with ageing or with disease. It's not just language learning that can increase your stock of cognitive reserve, by the way: higher education in general can do it too, as can a decent social life, and physical exercise.

'The hippocampal effect seems to be explained by time spent training rather than by proficiency, which lends some hope to those of us who work very hard but still perform so-so,' Mårtensson told me. In other words, just practising a foreign language, even if you don't get very good at it, may help you in the future. You may have found that a language you've not spoken for a long time can be reawoken if you finally have to use it – I've found this, to my relief, when visiting Japan after a long gap. Mårtensson says he'd bet on unused languages remaining dormant but viable in the brain, but adds: 'Brain structure probably benefits from you using it – not just for language – much like a regular muscle.' I really should practise my Japanese more often.

So learning foreign languages does seem to have a valuable neurological effect on the brain. The side-effects appear minor – bilinguals have a slightly smaller vocabulary in each language than monoglots, for example. There are other benefits. Bilinguals earn more than monoglots, for one thing.

There's another thing I've been hearing as I've met and spoken with polyglots. Some of them point to the way a foreign language has an effect on the way we think and act. Some bilinguals and polyglots say their character changes according to the language they speak. More flirtatious, say, if speaking Brazilian Portuguese. Melancholic in Russian. Pensive in French. Humble in Mexican Spanish. I'm certainly more indirect in Japanese than I am in English. But these may be national stereotypes that we bring to the language, or constraints imposed on us by the language, rather than actual changes to our personality.

Arguelles, for one, disagrees that his character changes in another language, but says your thought patterns do change, if only because of a language's cultural quirks. For example, Japanese requires different verb endings depending on the age and status of the person you're speaking to. I learned that it's just culturally better to give indirect answers in Japan. German influences the way you talk because the verb comes at the end of the sentence.

For Arguelles it might not change his personality, but perhaps that's because he's the introverted, bookish, rule-led type of polyglot. He's probably more polyliterate than polyglottal: a reader first, a speaker second. What about the more outgoing, empathic, social learners?

Richard Simcott, organiser of the polyglot conference, is on stage in Thessaloniki making jokes in various languages and gleefully threatening to sing a version of 'Let it Go' in twenty-five languages (I look this up on YouTube later and show it to my daughter, who is freaked out to see Elsa singing in Mandarin and Finnish and German and Catalan). Born in Chester on the English-Welsh border in 1977, Simcott is sometimes called Britain's most multilingual person, though he's uncomfortable with the title ('Have they checked everyone?'). But there's no doubt he's up there.

At the meeting in Greece he chatted happily in twenty-five to thirty languages. He speaks five languages every day at home (Macedonian, English, French, Spanish, German), uses up to fourteen professionally at work (he works in multilingual social media management) and has studied more than fifty. He seems to epitomise the social language learner. When he was a (monolingual) child he used to love imitating accents. 'When we went on holiday I would meet kids from different countries and it would always interest me that if I knew bits of the language I'd adapt the way I spoke, and I found I did that very naturally to fit in,' he says. 'And as a communication device to build friendships with people. So for me it's always been a very social thing.'

Simcott studied languages at university and at parties would go from group to group, switching languages. What he gets out of his languages seems different to someone like Arguelles, who is driven by intellectual curiosity about literature. But Simcott's personality, he says, doesn't change. The *perception* of his character might change because of the language he's using, but it's always him. 'More than a different personality, it's like wearing a different suit,' he says. 'It's like wearing different clothes.' So he might adapt how he asks for something in Dutch or German – more directly – so as not to be so maddeningly English about it. What drives him is the thrill of mutual understanding you get from social interaction.

He mentions during our chat that Icelandic is one of his favourite languages to speak, and I ask him to say something in it. God knows what he says but it sounds authentically Nordic. I can barely tell it's his voice. 'It's really breathy and wispy – there's something elvish about it, a delicate nature which I really like,' he says. Georgian, he continues, is his favourite written language, and when I look at an example I see what he means. The alphabet is beautiful. In fact it looks like a page of Tolkien's Elvish script: I think there's a theme here with Simcott. One of his favourite languages, though, is German. Not because of the language itself, which initially, like me and many others, he found unappealing, but because of the wonderful, welcoming people who speak it. I'll remember this later, when I meet some elderly British veterans of the Second World War and see that one of them is learning German.

Several of those attending the polyglot conference told me, half-jokingly, that they had obsessive-compulsive tendencies. One, a New Yorker, Ellen Jovin, said she set out at the relatively late age of forty to learn twenty languages. I think it's clear you need quite a drive to stick with that sort of challenge. She stays up late poring over Russian verb endings, then dreams about them. She's a self-described grammar freak.

So does Simcott think there are different types of learners? 'You get people at the conference who have difficulties in social situations, they're very good at languages, they love language in the same way. They describe themselves as introverts and extroverts, but the autism spectrum definitely fits in and features quite heavily in the community.'

Simcott says you find polyglots who by their own admission are socially awkward, don't feel comfortable in crowds, don't know what to say at the right moment, and sometimes feel they're not having the same emotional responses as other people. This doesn't mean you need to be on the autism spectrum, or have obsessive-compulsive tendencies, to become a polyglot. But it might help.

I've learned how thinking and working in another language is like using different software on the same operating system. It doesn't really change your underlying character but you might need to follow different rules. But it turns out that there is a deeper and more surprising effect of using a second language: it changes our morality. We become more utilitarian in our thinking. And if you ask people to perform risky tasks but constrain them to using their second language, they behave more sensibly than when doing the same tasks in their mother tongue.

Much of this work has been done by Boaz Keysar at the University of Chicago. Take the classic psychological test known as the trolley dilemma. This is an imaginary scenario where subjects are asked what they would do if they saw a runaway trolley (i.e. a train) about to hit a group of five people. You can do nothing, and let the people die, or you can throw a switch which diverts the trolley onto another track where only one person will die. Most people don't take long to conclude that it's morally justifiable to throw the switch and sacrifice one person in order to save five. The moral dilemma gets trickier in a second scenario, however: here, you are standing on a footbridge next to a large man. Again there is a runaway trolley heading for five people. This time, in order to save the five, you must physically push the man off the bridge and into the path of the trolley. It has the same effect – stopping the trolley from killing the five – but most people find the directness and associated emotional horror of the second scenario much more morally troubling.

You wouldn't expect your morality to change depending on whether you were posed the question in your native language or a foreign language that you understood, would you? But Keysar has found that it does.[7] He recruited native speakers of Korean, English and Spanish, and posed the dilemma to them in English, Spanish, Hebrew or French, according to their proficiency in the second language. In the footbridge dilemma, 18 per cent of the test subjects using their native language said they would push the man to his death, but in their second language 44 per cent said they'd do the deed. Participants were equally likely to sacrifice the single person in the switch version of the dilemma (80 per cent in their native language said they'd throw the switch compared to 81 per cent in a foreign language).

Keysar suggests that a foreign language gives you 'psychological distance' from problems – a bird's eye view of a problem can often help – and reduces the emotional response engendered by your mother tongue. So much for being more passionate when speaking Italian: if I was fluent in Italian my decision-making in that language would be dryer and more detached; what we'd characterise as being from the brain, not the heart.

Keysar has also shown that using a foreign language reduces your loss aversion. In other words it makes you more likely to accept a risky bet than in your native language.[8] In a review paper Keysar and colleagues note the evidence that decisions depend on whether the information is given in a native or a foreign tongue. They suggest that this has implications for the relationship between language and thought.[9]

Language plays a role in how we think and make decisions. 'What our studies show,' says Keysar, 'is that when you use your native language it plays an important role in your choices, it connects you more strongly to your emotions and thus affects your choice.' So the implications of the research go beyond the impact of using a foreign language. It tells us that most of the time when we use our native tongue it affects how we make decisions. 'We as researchers and people in general don't normally think that the fact we specifically used our native tongue is consequential in creating a certain decision bias.'

So it seems that, although our polyglots don't really change their personalities, they really do seem to change their moral thinking, and using a foreign language changes the way we make choices. The Dutch linguist and writer Gaston Dorren calls this the rationality effect: 'Speaking a second language not only makes you think harder about how you're saying things, but apparently also about what you're saying and even what you're doing. I think that's an amazing free gift.'

People need to know about this. Risks – such as those

associated with biotechnology or air travel — are perceived as lower when presented in a foreign language. *Capisce?*

I'd seen Michael Levi Harris in a short film he wrote and starred in, *The Hyperglot*, and I met him at the polyglot conference in Greece. He was there to talk about how the techniques he learned at drama school also applied to language learning. I was interested in this because it seems to me that a big part of effectively learning a language is to inhabit the personality of the relevant country. And indeed, don't you need to do that to properly get the rhythms and the character of the language? Learning a language is of course about far more than mere mimicry, but it involves observing, copying and empathising. So it does seem to have something in common with acting. Sure, it is both subjective and stereotyped to say that Italians are passionate and Germans are efficient, but for Harris, that's how he pictures the patterns of those languages playing out in his head, and it's how they come out when he speaks them. It helps him to get into character to imagine himself playing particular roles when he speaks a particular language. This was also fascinating to learn as I'd always wondered how they teach acting at drama school, and Harris, an American who has just graduated from the Guildhall School of Music and Drama in London, provided a glimpse.

Jacques Lecoq, Harris told us, was a French actor and theatre teacher who died in 1999. He is famous for his 'seven levels of tension' technique. Each level corresponds to the amount of nervous energy an actor puts into the performance. People use different names for each level but these are the ones Harris used. Next to each one I've put the languages that Harris associates with that level.

1. Exhausted (American English; Swiss French)
2. L'americano (Brazilian Portuguese; Australian English)
3. Neutral (German; Finnish)
4. Alert (British English; French)
5. Dramatic (Spanish; Greek)
6. Operatic (Italian; Hebrew)
7. Grecian/tragic (Chinese; Russian)

Harris used these to demonstrate, in English but simultaneously in American Sign Language, some lines from *Children of a Lesser God*.

He *is* an actor, so he *ought* to be good at it, but seeing the way he inhabited those different levels, and the way he expresses himself in different languages, I really warmed to the theory that languages have their own character and you take on some of those characteristics when you speak a particular language.

Maybe Arguelles' character doesn't change with different languages. With more malleable people though, their character changes according to the people they meet, even if they are speaking the same language. Actors are surely among the most empathetic of people, so an actor polyglot? It's no wonder Harris adopts such characteristics.

I considered empathy as a trait that should be given a chapter of its own in this book, by the way, but the problem is it's hard to measure. You know when someone is empathetic, and there is a kind of test for it: the emotional intelligence test, or EQ test. But although we like people who are empathetic, it's hardly a trait that enables you to find someone and say 'this person is the most empathetic in the world'. What's more, extreme empathy isn't necessarily a good thing. Yet a surfeit of empathy, an ability to be able to put yourself into someone else's shoes, is something that influences how good you are at several of the traits we'll meet in this book. 'We're wired to be empathetic, to put ourselves continuously in the position of other people and that makes us social,' says Cornelius Gross, of the European Molecular Biology Laboratory in Monterotondo, Italy. 'That makes us hurt when others hurt.'

If you have it, you'll likely be a social language learner, but if not, you'll be a lone studier. As I said, however, empathy is a fluffy and indistinct quality. It would be better if we had something a bit more definite to work with – and for that we need to know about the genetics of language.

Just over the river from where I live is the west London suburb of Brentford. There's nothing particularly notable about it, and for most people it is just a place you drive through on the way out of London to Heathrow. Built around the intersection of the rivers Brent and Thames, Brentford is most famous in our family for its football club, nicknamed the Bees – and soccer fans know that, uniquely in English football, Brentford's ground has a pub at each corner. But what they may not know is that in Brentford in the late 1980s the first step towards one of the greatest discoveries of human language evolution took place.

Elizabeth Augur, a teacher in a special needs unit at a Brentford primary school, was looking after seven members of a family now known in the genetics literature as the KE family. Augur discovered that a learning and speech disorder had been running through the family for three generations, and suspected that a genetic problem was at fault. This was long before the human genome project had sequenced our genes, so pinpointing the problem was an arduous process. Eventually, in 1998,[10] a team at the University of Oxford and at the Institute of Child Health in London traced the disorder to a region of chromosome 7 containing some seventy genes, and in 2001 they discovered the key gene that was damaged.[11] It was called fork-head box protein P2, or *FOXP2*; the press almost immediately dubbed it 'the language gene'.

That's not the case, but *FOXP2* did turn out to be a gene that has far-reaching effects. Not all of our 22,000-odd genes are turned on in all our cells, and some only operate at particular times in our development. *FOXP2*, however, operates in the foetal and the adult brain, and is turned on in the lungs, throat, gut and heart. It also conducts the activity of many other key genes. If it doesn't work properly, the outcome is like the conductor of an orchestra going awry: discordant sounds are produced. For example, members of the KE family who carry the mutated form of *FOXP2* are unable to pronounce consonants correctly, saying 'bu' for 'blue' and 'able' for 'table'.[12] They have speech apraxia: they can make the individual sounds, but the problems arise when they have to string them together in sequences, as needed for producing words and sentences.

We know that there are hundreds of genes involved in language.[13] It's what we'd expect from such a complex trait with so many elements — the muscles and nerves to shape the mouth and tongue and produce sounds, the control over breathing, the intellectual ability to learn grammatical rules and vocabulary. But *FOXP2* is one of the only genes we know that has a clear connection to language, so it has been studied intensively.

The gene is what geneticists call *conserved*. That means it has hardly changed over evolutionary time: it looks extremely similar in vertebrate species that are separated by millions of years of evolution.

When the human version of *FOXP2* was compared with that of chimps, Wolfgang Enard, then of the Max Planck Institute for Evolutionary Anthropology in Leipzig, Germany, found that there were differences in just two amino acids, out of a total of

715 in the gene.[14] There are just three differences between our version of the gene and that of mice, and eight between ours and that of birds. Conserved genes stay the same because they have profound and important effects. You don't mess with a winning formula.

It is sometimes said that the KE family mutation has reverted them to the chimp-type gene of *FOXP2*, but that's not the case. Instead they have a mutation that interferes with the gene's ability to control other genes. We know that the normal human version of *FOXP2* is intimately involved in our ability to learn new motor skills. Mice with mutant versions that damage the gene are less able to learn how to run on a tilted running wheel, for example.[15] And indeed, when mice were genetically engineered to carry the human version of *FOXP2*, their ability to learn a new task improved.[16]

Could it be that hyperpolyglots have a version of *FOXP2* that improves their learning ability still further? In the delightfully named Moody College of Communication at the University of Texas at Austin, Bharath Chandrasekaran is interested in just this.

Chandrasekaran knew that although it is highly conserved between species, there are subtle variations in *FOXP2* between people. Not like the serious mutations that cause rare and substantial disruption to language ability, as in the KE family, but in much more common changes in the DNA sequence that many people carry. At certain sites within each gene of the genome, people have common variations involving substitution of one nucleotide of DNA for another – an adenine (also called an A) instead of a guanine (G), for example.

Changes of this kind are called single nucleotide polymor- phisms (SNPs, pronounced 'snips') and they often have no effect on our phenotype – the body and behaviour built by our genes (if they did have a serious effect, they would be removed by natural selection). Chandrasekaran targeted one of the known variants of *FOXP2*, a SNP that is found in two versions, either an A or a G.

And since we inherit a copy of *FOXP2* from each parent, we can either have both As, both Gs, or an A and a G for this SNP.

Previous work had indicated that people with an AA-type polymorphism showed slightly higher activity in the prefrontal cortex when processing speech compared to people with a GG polymorphism. When you learn a skill you start by consciously practising it. This is called declarative learning. As you master the skill it becomes something you do without thinking, and this is when it is controlled by the procedural learning system.

Chandrasekaran wondered if the higher prefrontal cortex activity in AA-type people would slow their ability to learn, since the activity would slow the transition to the automatic, procedural learning system. He tested this by having volunteers with no knowledge of Mandarin Chinese, or any tonal language (one in which the meaning of a word changes according to the tone it is pronounced in), learn to categorise different tones in Mandarin. GG-type people were indeed better at learning the distinction.[17]

When we learn a second language we all start in the same way.[18] We begin with rules, and we can explain the strategies we're using to learn – for example, following a conjugation rule to add a particular verb ending. Chandrasekaran's work suggests that the more successful learners tend to shift to an implicit method of learning faster. In our example, this would mean the conjugation rules become accessed automatically. This allows the analytical, thinking part of your brain to do the other things necessary for learning a language, such as learning vocabulary.

'Language learning is like juggling while riding a bike,' Chandrasekaran says. If you can shunt some things into the procedural part of the brain, it helps with other things. 'That is, if you don't have to think about riding a bike because you have automatised that behaviour, you can focus on juggling. But if you have to think of every little thing that helps keep you on the bike, juggling is going to be next to impossible.'

Now, this is but one small polymorphism affecting, in a small way, a process (learning a language) that has a huge number of influencing factors. So even if we wanted to, say, engineer foetuses to carry the GG polymorphism, it would be unlikely on its own to reliably boost those individuals' language-learning ability. But Chandrasekaran has a better idea: create a behavioural intervention for those who struggle to learn a language. It would be a form of personalised medicine. People could get a routine genetic test to assess their version of *FOXP2*, and then have a schedule of learning devised for them. 'One thing to do is design language learning paradigms that help you with rules, but then slowly move to making things less rule-based and more implicit,' he says. 'Adults are wired differently and use different learning strategies. Current training approaches are not optimised to adult learning strategies.'

It's not a short cut to language proficiency, but then nothing is. What it might do – though given the complexity of the relationship between SNPs and language skills, many are sceptical – is offer a method to tailor your learning needs precisely to your genetic disposition.

Simon Fisher was part of the Oxford team who discovered *FOXP2*. Now professor of language and genetics at Radboud University in Nijmegen, the Netherlands, and director of the Max Planck Institute for Psycholinguistics, also in Nijmegen,

Fisher works on identifying and studying the functions of genes involved in language-related disorders, and is heading a project to look at the biological basis of exceptional language ability. He is less than convinced by Chandrasekaran's ideas on the GG polymorphism as he is troubled by an important technical detail. We saw above how you inherit a *FOXP2* gene from each parent, so the possible genotypes are AA, AG and GG. In a random sample, the numbers of people with each genotype should follow the pattern set out by an equation fundamental to genetics: the Hardy-Weinberg equilibrium. If they don't follow this pattern, then something odd might be going on – a mistake in genotyping, or a quirk of natural selection, for example. Chandrasekaran's team have acknowledged that their proportions don't fit Hardy-Weinberg, but for Fisher the mismatch, and the fact that individual SNPs have tiny effects on behaviour, throws doubt on the interpretation of the experiment.

Fisher's team are currently running their own experiments with *FOXP2* polymorphisms. For example, they have been assessing effects of *FOXP2* SNPs on language-related activity in the prefrontal cortex, measured with brain imaging in much larger groups of people than used for prior studies. So far, Fisher says, the search for a clear positive effect of one kind of polymorphism on language ability in the brain does not look promising. It did seem too neat, in a trait as complex as language, to have such a clear effect with such a relatively small genetic difference. We shall see.

On the other hand, since *FOXP2* is so involved in muscle control and articulation, Fisher wonders whether people who are at the extremes of articulation – rappers and beatboxers, for example – might have a particular variant, or combination of variants: 'There is an open question of whether proficiency in those domains might be influenced by genetic factors, or if everyone should be capable of mastering it, assuming they have enough training.'

Intuitively, in as much as that is worth anything, it feels like the capacity for language learning is partly genetic, but is something that most people need to work hard at. You need focus, which we'll examine in the next chapter.

4

FOCUS

Be true to the thought of the moment and avoid distraction.
Other than continuing to exert yourself, enter into nothing
else, but go to the extent of living single thought by single
thought.

Tsunetomo Yamamoto (c. 1710)

When I was in my twenties I was a member of a kendo club in
Japan. 'Kendo' literally means 'the way of the sword', and is
the training discipline developed by the samurai during Japan's
long feudal period. I loved it. At the club I was the only non-
Japanese, and during my whole time in Japan it was the only
place where no allowances were made for me being a foreigner.
Only there, for example, was I called Hooper-*kun* instead of
Hooper-*san*, '*kun*' being the familiar, informal suffix used for
younger men and boys. Kendo channels Zen philosophy, and
I loved being immersed in an element of the culture that was
very deeply Japanese. I learned about the difference between
what a European knight thought of his sword — as merely his
weapon — and what the samurai thought of it. In Japan the
sword was revered. Only the samurai were permitted to carry
one; it was their most treasured possession, their heirloom.

Once in training I leaned casually on my bamboo sword, called a *shinai*. My kendo *sensei* struck me hard on the legs with his, and reproached me. Your sword is your heart, he said. I should treat it with respect. I hadn't really grasped that my *shinai* was a stand-in for the samurai sword. I never leant on it again. Kendo is a state of mind as much as it is a form of physical training for fighting, and it was a privilege to learn from traditional, dedicated teachers.

One day we were visited by a highly respected master from another club. He was an old man, and we were given the honour of facing him in combat practice. He held his *shinai* carelessly, drooping almost, so it grazed the floor. When it was my turn he seemed to be looking away, lost in thought. I was facing him in the usual pose, *shinai* raised and poised. Yet when I went to attack him – it would require only a small, fast movement to strike his head with my sword – he moved in a way I couldn't afterwards adequately describe. With insane rapidity yet somehow also with apparent casualness, he raised his weapon and struck my helmet before I could complete my strike. Immediately afterwards he took on once more the appearance of a distracted old man. For the merest instant he had revealed his true self.

Okay, the anecdote may say more about my poor technique than it does about the old man's Yoda-like ability, but when he had left, my kendo *sensei* said the old master had extraordinary powers of concentration and reaction, honed over many decades of practice. He lived in the now. The old man encapsulates what we're going to explore in this chapter: the ability to focus, to concentrate and to react. Attention is a quality like the flame of a candle. It is ungraspable yet needs tending; it is always moving; it changes as you look at it. It is the 'now' we live in. This is going to sound like a Zen *koan*, but those who can master the now will be able to perform at a higher level. Those who can focus their minds can achieve great things. 'Concentration' is a faculty that exists on a spectrum. It may be exercised in short, intense periods, like the kendo master as he faced me, or it may be deployed in a sustained fashion, over the long term. We'll look at both forms, starting with the following example.

In 2004–5 Ellen MacArthur sailed 27,000 nautical miles around the world, non-stop, on her own. It took her 71 days, 14 hours, 18 minutes and 33 seconds, earning her the world record for a solo circumnavigation of the globe. She was twenty-nine. Many thought she wouldn't be able to break the record, which itself had knocked twenty days off the previous one and was considered safe for at least a decade. No doubt some of the scepticism voiced before her attempt was because she was a woman. But hers was an extraordinary triumph. In France she's been compared to Joan of Arc; in Britain she's been called the finest sailor the country has ever produced, and the first true heroine of the twenty-first century. I've come to meet her to find out how she did it. What does it take to focus so doggedly on something? How do you maintain concentration twenty-four hours a day, seven days a week, for two and a half months, on your own, with very little down time? It really does seem like a superhuman feat.

MacArthur is quite short, 5ft 2in, and I remember that the 75ft boat she broke the record with was designed and built especially for her small frame. The voyage was unimaginably tough. Her boat was a trimaran, because triple-hulled boats are faster, but they are also more unstable. MacArthur had come second in a round-the-world race, the Vendée Globe, in 2001. That race is between monohulled boats with keels: if they go over they will probably come back up. 'You flip a trimaran and it's all over and you're probably dead,' she says. 'At any point during the round-the-world, with a few exceptions, you could capsize that boat any time. You sleep with the ropes in your hands.' For most of the time, sleep, however, is an impossible luxury, and you have to get by on naps.

I ask her how she adapted to the stress. 'If you got on that boat and sailed around the world, you would have no choice,' she says. 'It's not about adjusting, it's about surviving. Literally. The boat is like an out-of-control Tube train, it's so violent.' Half the time, of course, you're sailing at night. You can never let go, literally or figuratively, for the whole trip. 'Mentally it is absolutely brutal. You get to the point where you feel you have nothing left and then something goes wrong. And you have no choice. No one's going to come and help you.' It sounds terrifying, I say. 'No,' she says. 'It's your home, it's your life. You do get used to it.'

Where does this drive come from, this ability to maintain such extraordinary focus, the resilience necessary to concentrate on the task? 'It all stems from having a goal. For me, from the age of four years old, I wanted to sail around the world.'

As a child she'd been on a boat with her aunt, and had immediately loved it. 'I thought this was the most amazing sense of freedom I'd ever felt, a boat can take you anywhere. I was so excited by that, I thought this was incredible: the boat had a little home in it, and you could travel with it.'

I find it extraordinary, instructive – and important to remember when we are dealing with them – that even very young children can have clear and burning passions.

From such a very young age, MacArthur says, she always had the idea in her head that someday, somehow, she would sail around the world. This was no phase she would grow out of.

She arranged her life so it would progress towards that goal, little by little. 'That really dictates the choices you make in life. At four you're tiny and your life is dictated for you, you don't have many choices. But as you develop you have these points where you have a choice.'

The first thing she needed to do was acquire a boat. The only way of doing that, not coming from a sailing family, was to save up. But she was a child, without pocket money. So she saved her birthday money and Christmas money, and even the money for her school lunch. She ate mashed potato and beans at school, every day, because it was the cheapest lunch, and saved the rest. 'I was totally focused. I could've had lunch but I didn't because I was going to buy the boat that I was going to sail around the world on. I didn't spend money on anything because I was saving for a boat. I'd even go down to the pub and not have a drink all night because I'd worked so hard to save the money.'

She is keen to assert her normality. 'I may have gone off and done some crazy stuff but I feel completely normal. I don't feel like I'm any different from anyone else, I just decided that was what I wanted to do.' She is uncomfortable with the idea that she is special, but I try and persuade her that most four-year-olds don't set out to do what she managed. 'You overcome barriers if you have this goal to drive towards,' is about as much as she will concede.

The other key point here is that you need that burning desire in the first place. It might strike some people like lightning at a young age, as it did MacArthur. But most of us have to work towards passion just as we'd set targets and work to a goal. We might get lucky and quickly stumble upon something we love and are good at, but most of us will need to shop around.

I love how Ellen MacArthur takes the pressure off the rest of us. 'If you really want to do something, don't think you can't,' she says, 'but most people don't really want to do something. And that's fine. That's brilliant. Life doesn't have to be about achieving amazing things. It just happened that I wanted to go sailing and sail around the world and then I went and made it happen.'

Identifying a goal and grasping it firmly seems vital for success, and for fulfilling our potential. We saw it with Hilary Mantel's long-term projects. It's something we'll see throughout this book, and it cuts across different traits and abilities. But what do we know about the brains of people who operate at high level in the now?

The din of tens of thousands of spectators. The roar of engines. The speed. The smells – the heat of the tarmac, gasoline, oil, rubber, hot metal, adrenalin. The anticipation. The money. Vast amounts of money. Formula 1 is a sport like no other. Hundreds of people work on each team, thousands of hours go in to preparing for each race, and on the day, it's about one man (it's still usually a man) driving one car.

On 25 June 2017, Lance Stroll, an eighteen-year-old Belgian-Canadian, was that man. He was behind the wheel of a Williams FW40 car capable of speeds in excess of 200mph, competing in the Azerbaijan Grand Prix. Stroll is the youngest driver currently racing in F1 and the son of a billionaire. He had been condemned the week before by Canadian driving legend Jacques Villeneuve as 'the worst rookie in F1 history'.[1] It's true he hadn't secured any points in the previous six races, but he had found something in front of his home fans at the Canadian Grand Prix, and finished in ninth place, earning two points. Then in Azerbaijan he drove a race of great maturity and skill, and finished third.

'It's a terrific result for Lance to become the youngest rookie in history to score a podium. He's had a brilliant weekend,' said Paddy Lowe, the chief technical officer at Williams. 'He's been faultless in every session, he's stayed out of trouble, didn't have any incidents and that carried into the race. He kept it clean, had good pace and managed the car and the tyres well.'[2]

I had no idea of the breadth of the skill set required to drive Formula 1 before I started researching this chapter. F1 drivers race for up to two hours. A Formula 1 circuit is between 4.3km and 6km (2.7 to 3.7 miles) long, and it takes about a minute and a half to complete one of the seventy or so laps. Drivers have to maintain intense concentration the entire time under conditions that would cause most of us to panic, spin off and wipe out. They need incredible reactions, of course, because they are moving so fast, keen spatial awareness and the ability to process everything that is coming in. Even professionals, on first driving a Formula 1 car, are amazed at how little time there is to think. Drivers need to be able to feel the car – it almost becomes an extension of their bodies. They need to be athletes to cope with the forces that go through their bodies. Racing exerts high g forces during turns, putting a lot of pressure on the head especially. This goes on for lap after lap, so they need endurance. They need race-craft: an ability to know when to overtake, how to use other cars to their advantage, how to bide their time. It is a kind of patience, but one conducted while driving at high speed. Technically the cars are not easy to drive: they are some of the most complex machines in the world. And there is, of course, real risk involved. A mistake at this speed could be life-threatening.

Very few of us will ever get to race an F1 car, but we all have shorter or longer periods in our lives where we are required to concentrate hard, when we need to focus. 'Sustained attention' is how psychologists talk about it. Perhaps if we could do this better, we could improve various aspects of our lives and our performance. The ability to maintain attention is linked to better outcomes in education and employment, and the converse – lapses in attention – blamed for a range of accidents.

I want to find out from those who represent the peak of potential for this trait, so I've come to the headquarters of the Williams F1 racing team in Oxfordshire, to talk to Stroll and Luca Baldisserri, formerly sporting director at Ferrari and now mentor to Stroll at Williams.

Motor racing has an incredibly high rate of attrition. Partly because it's an expensive sport, but also because intense natural selection culls drivers at all stages, from karting, through Formulas 4 and 3, to Formula 1. It is perhaps the most elite of sports, and Stroll is elite in two senses. First, his father has given $80 million to Williams. This has led to claims that Stroll has bought his place in the car, but Lowe and the team hope the performance at Azerbaijan will start to lay those criticisms to rest. Second, to drive in Formula 1 you first have to earn an FIA (International Automobile Federation) Super Licence, and Stroll was also Formula 3 champion in 2016. Sure, $80 million doesn't go amiss, but, say Williams, Stroll has earned his place on merit.

I'm entering an exclusive world here. I clear security and steer my jalopy to the vast car park, which is thankfully not full of high-performance vehicles, then meet my chaperone for the visit. I am not allowed to see the 2017 car Williams are racing this year, but I get up close to the 2016 model. I've never been a particularly ardent fan of motor racing, but there's no doubt that this is a beautiful piece of engineering. The steering wheel is baffling – it looks like all the controls in an aircraft cockpit have been condensed on to it. I was hoping to have a go on a driving simulator, but I'm not allowed even to see it. Both the 2017 car and the simulator are fitted with proprietorial, sensitive technology that the Williams engineers do not want leaking. The simulators, by the way, cost many millions of dollars. They are set up by a team of dedicated engineers who monitor parameters such as tyre contact, downforce and engine performance during the simulated race. Drivers sit in the same cockpit as they do when they drive for real, wearing a helmet. It's as close as you can get to racing an actual F1 car. Most teenage boys have PlayStations; Lance Stroll has a full F1 simulator of his own in his apartment in Geneva.

Stroll is tall, taller than I imagine is optimum for a driver, dressed in jeans and black T-shirt. He's relaxed. He says he lives in the moment, as any eighteen-year-old son of a billionaire ought to. But he is grounded, and there's nothing of the playboy racing driver about him. He seems more astute and mature than I was at that age, that's for sure, but then he's had media training. When he says he lives in the moment he's referring to when he's driving. Far from being distracted by all the noise and vibration and the mayhem of the track, he relishes it. It's when he's in the cockpit that he performs best. 'When I'm behind the wheel and I have my helmet on there's no distraction, there's nothing bothering me, nothing getting in my way, it's the real me,' he says. 'I'm very competitive, I'm an attacker and I love speed and I've always loved motor sport.'

I speak to James Hewitt about what it takes to perform at Formula 1. Hewitt is head of science and innovation at Hinsta Performance, a company who work with athletes and professionals, including F1 drivers, to optimise their capabilities. Previous clients have included former F1 world champions Sebastian Vettel, Mika Häkkinen and Nico Rosberg. Hewitt says there is always a mixture of intrinsic and extrinsic motivation in drivers.

The rewards of winning are huge, and the young ones especially feel the pressure to justify the massive investment their supporters have made in them. But there is also huge hedonistic pleasure just in driving at great speed. For most of us, driving so fast would cause stress under which we would buckle. 'But for these high performers,' Hewitt says, 'they interpret that stress as pleasurable.' An average person's best performance comes with a moderate level of arousal, but in motor sport the drivers' optimum performance comes at a higher level. 'The world is coming at you at 300km an hour, it's also very loud, there's a lot of vibration, so your body is being highly stimulated,' says Hewitt. 'The most successful drivers I've worked with, you ask them what they feel and they love it, they absolutely love it.'

Stroll's love of it comes through when I ask him what he gets out of it. 'I get to be myself on the race track,' he says. 'In the car, I'm not doing it for anyone, it's just me, the car and the track, and that's my passion. I feel alive and I feel that's my drug.'

Various experiments have been conducted to test the effect of this kind of motivation on our powers of concentration. Michael Esterman is co-founder of the Boston Attention and Learning Lab, and a psychologist at Boston University. 'The science shows that when people are motivated, either intrinsically, i.e., they love it; or extrinsically, i.e., they will get a prize,' he says, 'they are better able to maintain consistent brain activity, and maintain readiness for the unexpected.' Motivation means this consistency doesn't fall off over time.

Esterman's team has performed several experiments to test this. In one, participants were shown a random sequence of photographs of cities and mountain scenes, one every 800 milliseconds, while in an fMRI brain scanner. They had to press a button when they saw a city scene (which occurred 90 per cent of the time) and avoid pressing the button when a mountain scene appeared (the remaining 10 per cent). Sometimes they took part in trials which were rewarded. In that case, participants earned 1 cent for each city scene they responded to, and 10 cents for not responding to a mountain scene. They were also penalised for getting it wrong. Other trials had no reward or penalty. The results of their brain activity showed that without the motivation of reward, the participants acted as 'cognitive misers': they didn't bother engaging the brain's attentional resources until their performance had dipped. Until, in other words, they had dropped out of the zone. When they were motivated by reward, however, the participants were 'cognitive investors', happy to engage their brain and concentrate in order to stay focused on the task.[3]

'Motivation,' Esterman says, 'plays a big role in maintaining optimal attention and focus.' When people are in the zone, they use attention brain regions more efficiently, he explains. 'There is greater connectivity, communication and information transfer in the brain, and they process sensory and visual information with higher fidelity.'

This is what's happening when Stroll is driving. It's almost as if everything in his head is working faster, so perhaps to him driving at 200mph is like me driving at 100. Of course, racing a Formula 1 car is a different proposition to pressing a button in response to photographs. But the experiments support the idea that a reward helps motivate focus. 'Researchers used to believe that attention was a finite resource that could be drained and then had to be replenished,' says Hewitt. 'But it's more complex than that, and it has a deep relationship with our motivation.' When reward and motivation are added to the already world-class talent demonstrated by drivers who make it through to F1, you get exceptional results. 'Having that incredible prize on offer coupled with the selection pro-cess results in this incredible sustained attention that we see,' says Hewitt. Their attention is such that F1 drivers typically have an uncanny ability to remember what's going on in their environment in high definition. Hewitt calls it a superpower, and gives the example of a driver in a simulator. 'If you pause the simulator, the driver would be able to describe the cues they're looking at.' For example, drivers will use buildings as cues for braking, but they'll also be able to tell you about what's happening on the track: the location of their competi-tors, as well as data on how the car is performing. Hewitt says an F1 driver's working memory can hold more discrete items than can that of a normal, untrained driver. F1 these days is very strict about allowing races to continue if weather condi-tions get too bad, but not so long ago drivers sometimes raced in torrential rain, with visibility essentially zero. 'One driver said if you're driving in the rain at 300km/h you essentially have to determine your braking point and when to turn on intuition and experience and what you remember,' explains Hewitt. The driver in question described being on the straight and counting. When he reached a particular number – one one thousand, two one thousand, three one thousand, say – he would brake and turn.

As to why Stroll loves it so much – why he feels more himself when he's driving – that's because he has achieved the state of flow, and flow is a desirable place to be. This is the concept introduced by Hungarian psychologist Mihaly Csikszentmihalyi. Flow – what Esterman calls being in the zone – is when you are absorbed by a challenging task such that you are completely concentrating on it. The task isn't so chal-lenging that you panic and are unable to perform, but neither is it so easy that your skill is too great to make it interesting. This is the state F1 drivers enter for two hours when they race. I have a small sense of what this must be like. When I'm snowboarding sometimes I can get into a rhythm of carving turns that is incredibly meditative and transportive. Even cycling to work through London traffic can be pleasurable because my mind is diverted from the normal worries.

Stroll explains what is most difficult, mentally, about racing: 'The hardest thing is going a bit beyond your comfort zone in the qualifying lap. To get that extra two-tenths of a second. You know you can get them. But the risk of completely losing it is high.'

I ask if there's fear, already knowing that he will say there isn't. Hewitt told me that typically the young drivers won't even think about the dangers, or if they do they very quickly pass on and dismiss them.

'It's not scary, it's risky,' says Stroll. But the risk he's talking about is not the possibility of real bodily harm, it's about not making the manoeuvre you intended, and losing time. 'It's easy to say in qualifying "this is the time I can do" but you have to push yourself to get there. It applies in tennis. You want to hit a shot across court. It's risky but can you do it. That's what's enjoyable.'

If it works, you get a great lap in, and that's the reward. 'Your body takes over and you're completely free and it's overwhelming,' he says. 'It's quite a feeling.'

Stroll's current driving coach Luca Baldisserri was for many years a race engineer at Ferrari, where he worked with Michael Schumacher and Kimi Räikkönen before developing Ferrari's young-driver programme. He sees many young children who can drive, he says, but what he looks for at this early stage — what he considers most promising in a kid — is not driving ability. It is whether they have a goal. It's what we saw with Ellen MacArthur. Young people, he explains, are very easy to distract, they easily lose concentration. 'So the main thing you're looking for with a child of thirteen is if he or she has a target.'

He met Stroll when the latter was twelve, and could see he was special. Stroll was winning races in karting in North America, but Baldisserri wanted to know if his target was to become a Formula 1 driver.

That was the case with Stroll, even as a child. Like all racing drivers, he was and is super competitive. I ask him about his goals. 'You have to live in the moment,' he says, 'but my goal is to be Formula 1 world champion and that's what I work towards.' I suppose you might as well aim for the top. 'I want to be the best possible driver I can be,' Stroll says. 'I take it day by day. And win the day every day. Win the day. That's what your focus has to be towards.'

'Win the day' sounds like a line from a psychology-of-performance manual if ever I heard one, but it develops a mindset of progression, and that's Baldisserri's job. 'We build intermediate targets, moving up gradually, moving step by step,' he says.

He breaks down the training programme for a driver into three main components: physical, technical and mental. He sees the driver as an athlete, who has to train to be physically prepared, while equipping himself technically to drive the car. 'You have to deal with the weather, the temperature. You have to understand how the tyres are behaving.'

As well as the ability to maintain focus, you need mental strength.

'It's different to team sports like football. A driver is alone when he's driving. The family, sponsors, fans, all are watching.

Then the drivers have to deal with all the comments.' Stroll has had a lot of negative comments. 'It's in your face when you look at your phone,' says Baldisserri.

'The drivers need to stay concentrated on their objective. You need to create a target for them and help them cope with stress.'

Just as we say some singers are naturally gifted, the same is said of some racing drivers. Lewis Hamilton in particular is often spoken of as having huge natural talent. Baldisserri believes real talent can be seen in the way that a driver judges the battle during a race, assessing how to overtake or protect himself from being overtaken. He needs, too, to have a vision of the race. 'You need to have the car in your hand, completely. That's the main difference between a talented driver and an average driver. The technique of driving you can learn.'

He illustrates this by explaining what he means by having the car in your hand. There are sensors in the car that log in detail all the movements of the vehicle. They can measure the delay between when the car starts to move and when the driver makes a correction. 'And if you see shorter delays it means the driver has a lot of feeling. And this is a way of judging if a driver will be able to cope with cars that are faster.'

We've seen the power of long-term goals, with Ellen MacArthur. And we've seen how reward and motivation can help Formula 1 drivers to maintain intense focus. But what's happening in the brain? To understand that, we need to look at people who are experts at inhabiting the now. We need to find them, and put them in fMRI machines. Fortunately, there are many such people, and they've been studied extensively.

When Yi-Yuan Tang was six, he was a schoolboy in Dalian, a relaxed seaside city in Liaoning Province, north-east China. It was here he started contemplative practice, a cousin of meditation, during which you reflect upon your behaviour, and train yourself to enter a quiet, thoughtful, peaceful state.

As a boy he started running, and found he was good at long distances, especially 3000, 5000 and 10,000 metres. 'I won all these competitions in my middle and high school years,' he says.

Children are introduced to meditative practice at an early age in China. Tang realised that the practice of narrow-focus meditation – where you focus inwardly, usually on your breathing – was helping his running. Then he started practising open-focus meditation. This is often performed with a view to optimising the level of attention you bring to bear on something. Imagine being Lance Stroll driving in that Azerbaijan race. You don't want to focus too much on what you're doing, because that could lead to over-thinking. At an elite level, this will be to the detriment of your performance. At the same time, of course, you need to be completely focused on the task at hand. This is what mindfulness and meditative practitioners call the balanced attention state, but is perhaps best understood as being in the state of flow. Decisions are made swiftly, correctly and often without conscious thought.

Tang found that meditative practice and running had much in common. More than that, they interacted, so as he became more adept at meditation, he found his running benefited. 'It significantly improved my performance because this effortless attention and action reduced my stress and facilitated my flow state during running. In my opinion, a peak performer should use both open- and narrow-focus strategy in meditation or sport.'

Tang maintained his interest in the training of the body and the mind, and the interaction between the two. In the 1990s, borrowing from traditional Chinese practice, he developed a form of mindfulness meditation called integrative body–mind training (IBMT). It emphasises acknowledging internal and external distraction, for example, awareness of a sore back, or noticing the noise of people talking around you. It is, he says, about accepting these things with equanimity – a concept we'll come back to at the end of the book. But Tang also became a scientist. He currently holds the Presidential Endowed Chair in Neuroscience at Texas Tech University, Lubbock. And he is a professor in both Texas Tech's department of psychological sciences and in the department of internal medicine at the university's Health Science Center.

Tang's work has done much to demonstrate scientifically the effects of meditative practice on the brain. In 2015, with Michael Posner at the University of Oregon, and Britta Hölzel at the Technical University of Munich, Germany, Tang published a review of the evidence in the prestigious journal *Nature Reviews Neuroscience*. The trio concluded that more than twenty years of research into meditation supports the idea that it is beneficial for physical and mental health, and that it improves cognitive performance.[4] It improves brain power, basically.

For example, Joshua Grant at the University of Montreal in Canada scanned the brains of Zen practitioners who had racked up more than a thousand hours of practice. These seasoned meditators showed less activity in a few areas of the brain than non-meditators: in the prefrontal cortex, the amygdala and the hippocampus.[5] These areas are respectively concerned with (among other things) awareness of pain, the processing of emotions such as fear, and memory storage. But some parts of the brain that process pain were thicker in the meditators.[6] There's no contradiction here: meditators process the pain, but let it bother them less.

Another study shows how meditators may have a better connection with their subconscious. It follows a famous experiment first carried out in 1983 that suggested we had no free will. The late Benjamin Libet, then a physiologist at the University of California, San Francisco, measured brain activity as volunteers first decided to press a button, and then actually pressed it. The shock result was that the parts of the brain that control movement become active before the volunteers think they first make the decision to move their finger. This probably just means that our conscious awareness of making a decision lags behind a bit, not that we don't have the ability to make decisions at all and don't have free will. Still, in 2016 Peter Lush at the University of Sussex repeated the experiment, but this time using people who regularly meditate. There was a longer gap between when the meditators felt they decided to press the button, and when their finger moved to press it, than in non-meditators. Lush suggests this means that if you meditate you have heightened awareness of your brain activity.[7] You could say it means you know yourself better.

On a whim, once, I attended a four-day residential course on Zen Buddhism, held at Enkakuji temple in Kamakura, Japan. The temple was founded in 1282 and we stayed in a wooden building in the grounds that was almost as old. At four each morning we were roused by the monks to do morning meditation. It was March, still chilly, and I was sat facing an open window. A complete novice, I had no idea what we were supposed to 'do', so I sat there cross-legged for hours and my mind wandered. Occasionally a monk farted and I tried not to giggle. I remember suddenly noticing that my thoughts were flowing through my mind and I was watching them, like I was watching fish in a stream. It was like hypnagogia, the weird state between being asleep and awake (we'll come back to this in Chapter 10). My other vivid memory from that morning has nothing to do with meditation but everything to do with Japan. We sat for hours and gradually dawn broke, and I could start to make out the branches of the plum trees in the temple garden, and then the plum blossom. If not for the flatulent monks it could've been a scene from a Yukio Mishima novel, but I cherish the memory all the same.

One of Tang's studies shows you don't have to be a monk with a thousand hours of experience to get the benefits. Tang and colleagues recruited eighty-six Chinese undergraduates and randomly assigned them to two groups. Half undertook five days of his integrative body–mind training. They spent twenty minutes per day on the exercises. The other half spent the same amount of time in relaxation training, which meant they learned how to voluntarily and progressively relax their muscles. All the students were assessed before and after the five days using the Attention Network Test, a computerised assessment that measures alertness and the ability to resolve conflict. They were also scored for their emotional state using the Profile of Mood States test. The results indicated that the students who had taken the IBMT meditation classes showed greater improvement in the Attention Network Test, and exhibited lower levels of anxiety, depression, anger and fatigue, and higher vigour on the Profile of Mood States scale.[8] The IBMT classes were run by experienced coaches, but still, it cuts against the assumption that you need to meditate for years before you get the benefits.[9]

Meditative practice leads to changes in the very structure of the brain.[10] Two areas of the brain known to be key to our ability to focus attention, the anterior cingulate cortex (ACC) and the insula, a deep fold in the cerebral cortex, both grow in people who meditate. These regions, along with parts of the front midline of the brain called the anterior cingulate gyrus, are activated during cognitive tasks. The ACC, for example, aids in the maintenance of focus by preventing other systems in the brain from barging in and demanding attention. When we are performing tasks that have been practised over and over, such as adjusting the sails on a trimaran or changing gears in a racing car, the autonomic nervous system plays a big part in carrying them out. That's the part of the nervous system that acts automatically, performing functions such as regulating the heart rate and digestion. When we are in an effortless state of flow this occurs below the level of conscious awareness, and the ACC and the insula together help the autonomic nervous system achieve it.[11]

'We have examined training in mindfulness meditation, which involves holding attention rigidly fixed in the present and provides a task of high concentration,' says Michael Posner, referring to the narrow-focus kind of meditative practice. 'We have shown that such training increases activation of the ventral ACC and changes white matter pathways surrounding the ACC.'

How does this relate to elite athletes and people at the peak of performance, who have to make decisions fast and under pressure? Perhaps their brains work differently to those of average performers. 'One of our working assumptions is that intentions can be unconscious, and that decisions can be made unconsciously too,' says Peter Lush. 'However, once intentions become conscious, they probably become available to other processes.' When this happens, he says, it becomes a hindrance to peak performance.

Lush's intuitive understanding of flow is that it works not by engaging with our intentions but rather by allowing us to be a non-judgmental observer of them. For Formula 1 racing, or round-the-world sailing, that would involve sustained concentration on the present moment, which is similar to some aspects of mindfulness, and equates to Tang's narrow focus, but it also requires open focus, a switch to a global view, and the ability to return to dynamic flow.

It might be that the brains of Lance Stroll and Ellen MacArthur are naturally effective, poised to act as conduits to provide high levels of focus. 'Your hypothesis might be correct although we don't have clear supporting evidence yet,' says Tang. But probably, too, their relentless pursuit of a clear goal, and their long years of repetitive training, their expertise in their chosen sport, while not overtly influenced by mindfulness and meditation, had a similar effect to meditative practice and structurally changed their brains. Tang agrees: 'Activities such as exercise by an athlete can also lead to brain plasticity.' He is now able to meditate himself into a state of 'static flow', in which his mind is beautifully focused, and transfer this to a state of 'dynamic flow' when he runs. 'In this situation, my body and mind work together perfectly and optimise my performance.' The good news from all this is that the benefits are available to anyone, and it can be easily practised at home.

Now, when I think back to how the old man in my kendo club so easily beat me, I feel I understand much better. It's not that I'm still smarting because I thought I was so good, but that now I understand how he was so good. Kendo training is strongly influenced by Zen meditative practice, and we've seen how the brains of Zen practitioners change as a result of their training. On top of this there's what I've learned from Tang about physical training, the autonomic nervous system and how those skilled in meditation and sport are able to transition between a static and a dynamic state of flow. The old master had spent decades practising kendo, and even standing still I guess he was holding his mind in a perfect state, in the moment, poised but in flow, ready.

Part II

DOING

5

BRAVERY

A hero is no braver than an ordinary man, but he is braver
five minutes longer.

Ralph Waldo Emerson

Dave Henson is a powerful man: lantern jaw, massive torso; a
former army captain. Big Dave Henson, as he's known. You'd
want him on your team. He was a bomb disposal officer in the
British Army Royal Engineers. It was his job to look for improvised explosive devices (IEDs) planted by the Taliban.

US military data released by Wikileaks shows that in the
aftermath of the US-led invasion of Afghanistan in 2001, IEDs
gradually became the Taliban's go-to weapon. Between 2004
and 2009 Taliban fighters planted more than 16,000 homemade
bombs across the country, the numbers increasing each year of
the conflict. Detonated remotely, or by timer, or by a trip wire
or pressure plate, IEDs killed many hundreds of civilians, and
soon became the biggest killer of coalition forces. As well as the
hundreds of deaths, thousands of soldiers lost limbs in surprise
blasts. People joked darkly that at least the Paralympics teams
would benefit.

War is hell, yes, but the threat from IEDs, the danger they

posed, and the pall of fear they cast across both the armed forces and the civilian population created a particularly tense atmosphere in the country. By 2014 there had been more than 70,000 cases of post-traumatic stress disorder (PTSD) in the US Army alone. Whatever you think of the political decision to go to war, for a soldier to choose to serve in Afghanistan to look for IEDs takes a particular kind of courage.

I've chosen Dave Henson's story to examine what bravery is, but it could equally well feature in the chapters on endurance, resilience and even happiness. His story, as an aside, also works to illustrate the peculiar, fateful path that our lives can take.

Henson grew up in Southampton, on the south coast of England. He studied mechanical engineering at the University of Hertfordshire, and as part of that degree you have to work for a year in industry. Since the army offered placements, he did a year with the Royal Engineers. When he had to decide a subject for his dissertation, one of the garrison engineers suggested doing something on injured soldiers. 'This was 2006, things were kicking off in Afghanistan, and people were starting to come back with no legs,' Henson says. So his project was about getting disabled people back into sport – specifically, about amputees taking up go-karting. 'I had a wheelchair in my student flat I went around in. I figured I wouldn't know what it was like to get into a go-kart from a wheelchair if I hadn't tried it, so I went around in this wheel-chair – it was an absolute pain in the arse. But as a result I got quite good in a wheelchair.'

That skill would come in useful. After graduating, Henson stayed with the army, going to the Royal Military Academy Sandhurst, the army's officer-training college. Having passed out from Sandhurst he joined 22 Engineer Regiment, and then a position came up in the Royal Engineers as a search advisor in the Explosive Ordnance Disposal (EOD) unit. 'When the job came up it did have the element of risk but I saw it as something intrinsically interesting,' he says. The element of risk, he tells me later, with as much drama as if forecasting the chance of rain next week, is that there's a one in six chance that someone on a team will be injured or killed on a tour. One in six. How many people accept those kinds of odds at work?

Henson and his team trained extensively for various casualty scenarios, as they are called in the army. But he saw first hand what blast injuries can do to a person. This was no scenario, it was real-life blood and gore. 'An Afghan laying IEDs triggered his own device and was brought into our base,' Henson recalls. 'So seeing his . . . remains . . . he was still alive but didn't survive. That was quite horrendous.'

In February 2011 Henson was in Helmand Province, tasked with making areas safe for families displaced by the Taliban to return to their homes. Methodically checking and clearing the ground of bombs is often protracted and mundane work, but 13 February was more memorable. Henson's EOD unit was in the south of Nad-e Ali district, an agricultural region about 100 kilometres west of Kandahar. It had been raining for a couple of weeks, and this was the first dry day they'd had in a long time. The team cleared the first compound and moved on to the second. Tensions were high, as four days earlier in the north of the district two soldiers from the Parachute Regiment had been killed in a gun battle.

'I crossed the outer compound to get visual contact with infantry soldiers guarding us, walked back and that was it,' he recalls. 'There was no click, nothing like that, I just remember landing on the floor, and sitting up. It felt like someone had hit me with a spade, my head was ringing, I had tunnel vision. I looked down at my legs. They were attached but were hanging on by strips of skin, with bones poking out.' His feet were still in their shoes, which was nice, he says. 'I just remember screaming and backing away against a wall. One of the soldiers came into view and that shook me into level-headed thinking and suddenly it was back to reality.'

He's a powerful man, as I said, but I didn't mention that his legs end just above where his knees ought to be. His Twitter handle is @leglessBDH. When I meet him it's a cold winter day and he's wearing steel prosthetic legs with latex pink bare feet. He doesn't need to wear shoes indoors as the rubber feet give a good grip. Ridiculously, I have to stop myself asking if his feet aren't cold in this weather.

Most injuries such as Henson's result in death, but on this occasion the helicopter arrived within twenty minutes of him being blown up. He spent the time, jabbed with morphine, smoking cigarettes with his team, as if whiling away the day. Nevertheless he was in considerable pain. 'It felt like someone had parked a car on my legs. It was an unmoving pressure pain that just wouldn't go away.' Henson managed to distract him-self from the pain until the helicopter came, and he was on the operating table at Camp Bastion after thirty-seven minutes. There's no doubt that this rapid reaction saved his life. Back in the UK by the next day, he had his right leg amputated above the knee, and the left through the knee.

'I knew it was a risky job,' he says now of his choice of occu-pation. 'I'd assessed the possibilities and even the likelihood of coming back with some kind of injury like this so it wasn't altogether surprising that it happened. So it wasn't necessarily a shock.'

Speak to fire and police officers and you hear a similar thing: dealing with danger becomes just part of the job. As Henson explains, you don't allow yourself to be consumed by fear of what might happen, else you'd never get anything done. There is an ever-present underlying tension as the potential for disaster is always there, but you learn to live with it.

What I find extraordinary about people like Henson is that they can go into a job such as this in the full knowledge that something bad is quite likely to happen. They know the risks and do it anyway.

Bravery comes in different forms but bomb disposal seems to require a particular kind. Operatives in this field are brave in a way that marks them out even from other members of the military. They knowingly choose a career that exposes them to a constant and particularly insidious kind of danger, invisible and potentially lethal. One of the commanding officers of the engineers said as much when the British Army introduced a new badge to be worn on the uniforms of EOD personnel: 'To do this all day, every day, for six months demands a certain kind of mettle – a persistent courage.'[1]

Henson is the epitome of the modest British officer. They view the use of understatement as a kind of competition: sure, there's an 'element of risk'. At one point Henson refers to his double amputation as 'just a flesh wound'. But he is serious about what he regards as bravery. He says that if what he did was brave, it was a collective bravery. 'It's not me going out on a lonely walk on my own to find bombs, it was me and a team of people – and that made all the difference.' It's similar to other high-risk military roles, he says, in that you're rarely doing them on your own. You don't have to worry so much about the immediate risks to yourself because you've built up a trust and a cohesion and a belief that you'll get through it together.

He cites the army slogan: the team works. 'It means immediate risks or the extent to which you think of them are lessened,' he says. 'If you dwell too much on how much this could hurt or what the potential consequences are you'd never get out the door. But together as a team you can get through it.'

I begin to see how bravery can be collectively bolstered by being in a team. But what is the motivation driving people to do these things? Where do they get the guts to do what they do? I don't think I could do it, and I'll side with Falstaff on this. His rejection of honour is effectively the same as rejecting bravery:

Can honour set to a leg? No: or an arm? No: or take away
the grief of a wound? no.
Honour hath no skill in surgery, then? no. What is
honour? a word. What is in that word honour? what
is that honour? air. A trim reckoning!

Fortunately for the army, there are many people who think differently. Julie Carpenter interviewed twenty-three US Explosive Ordnance Disposal personnel (all but one of them men) for her 2013 doctorate at the University of Washington, Seattle. 'One thing that I discovered strongly attracted people to EOD,' she explains, 'was the fact they believe their role is "helping" as opposed to harming, because their work is to render safe unexploded ordnance.' This is similar to what Henson told me about his motivation.

Teamwork is absolutely key, Carpenter says. 'EOD are regarded as a little rebellious in spirit; they are viewed as confident, smart, and close-knit even compared to other military.'

You, like me, may have been reminded of the movie *The Hurt Locker* during all this, but if so, here comes an admonishment. That film gave an inaccurate Hollywood version of bomb disposal, says Carpenter, focusing as it did on a renegade main character who was not a team player. Real EOD work, both

Carpenter and Henson emphasise, relies on close communication and teamwork. 'EOD have incredibly strong teamwork and they exhibit this every day, but they enjoy their reputation as rebels within the military,' Carpenter says. 'EOD must have not only the willingness to work well in a team, but strong abilities to communicate as it is essential for the team members, who are in almost constant communication during a mission.'

According to Henson there are various reasons why you sign up to serve but the baseline is always to make a difference. His team, he explains, were conducting work in an area that had seen intense fighting. In an effort to further the political aim to create peace and stability, their task was to clear compounds so farmers could return to their land.

He takes pride in saying he was a Royal Engineer Explosive Ordnance Disposal officer. 'It was our job to look for bombs, and bombs were what was killing troops. Of course there's also a status to it, an ego attachment to it.'

Other sources of motivation inevitably include the excitement. Henson admits he joined the army to experience risk. 'Perhaps it was adrenalin, a rite of passage, a need to prove myself – there are a whole load of reasons why people join the military.' When the job came up he saw it as something intrinsically interesting, but was also attracted by the element of risk. 'The risk added to the enjoyment factor. The adrenalin rush you get from it is part of the buzz.' Youthful bravado is certainly a factor: Henson says these days he's not like that.

As an evolutionary biologist my first thought when I encounter a behaviour is often to wonder how it might have evolved. In the case of bravery there are a number of possibilities corresponding to the various forms that courage can take: brave acts protect your family and friends, perhaps rescue loved ones and, in ancestral times, provide food. Brave acts may also demonstrate credentials for leadership and perhaps suitability for a partner. Bravery, all told, is an impressive and attractive trait.

Falstaff will have none of it. And I won't argue with one of the greatest wits in literature except to say that Falstaff isn't renowned for his sex appeal. Bravery may be just a word and have no skill in surgery, but people who exhibit bravery are nevertheless considered more attractive by the opposite sex. Firemen certainly enjoy this reputation. Henson isn't convinced:

'I don't reckon the lads got any extra action because they were in EOD.' But, he allows, 'If you say you're an EOD operator or bomb disposal operator it certainly gets you into dinner parties.'

Thanks to the strong relationships forged in the military, people are willing to do almost anything to help another person. How do we explain that, from an evolutionary point of view? Laurent Lehmann, of the department of ecology and evolution at the University of Lausanne in Switzerland is, like me, an evolutionary biologist. Lehmann and his colleagues build mathematical models to explore how behaviours such as bravery, altruism, leadership and despotism may evolve.

The simple explanation of bravery, in evolutionary terms, is that we are predisposed to take risks when our family are in danger. Lehmann puts it more technically: if bravery is costly for an individual and puts him or her at risk to an extent that its action can never be repaid during the individual's lifetime, then the only way it can evolve and be hard-wired is by kin selection. This, in other words, is when you do something that has no direct benefit to you but does benefit your family. Examples in animals include birds that spend time and energy feeding the chicks of their siblings.

'The platoon is like a family as it creates the same physical proximity and maybe the same kind of sensory inputs – it creates "bands of brothers",' says Lehmann. 'So maybe the action of soldiers in the environment created by the military leads them to take actions as if they were helping their kin.'

For thousands of years, military leaders have recognised the psychological value of creating tightly knit groups of men. Perhaps they've been tapping into deep evolutionary behaviours, too. Once bravery has arisen in people by this process of kin selection, it can be co-opted by the military in ways that mimic the conditions under which it evolved. While most of us feel we wouldn't be able to exhibit such persistent bravery, the evolutionary argument and the team bonding dynamic goes some way to explaining it. It also suggests bravery can be learned, to some extent, or, in the case of the army, instilled. It can also lead to extreme cases of bravery.

The protocols of the team and the intense training he and his colleagues had been through helped Henson stay calm during the wait for the helicopter. It's the team, the camaraderie, that makes the whole endeavour possible. 'Genuinely in these situations you're next to people you're desperately fond of,' Henson says. 'It makes you do strange things. It's hard-wired into us to do these things for other people.'

Indeed, in those situations people do strange and remarkable things. Peter Singer, a warfare expert at the New America think tank based in Washington, DC, has described the case of a US soldier in Iraq – also an EOD soldier, as it happens – who ran 50 metres under machine-gun fire to rescue a team member. Remarkably brave in itself, but then consider that the team member was a robot that had been knocked down. Why would you take such a risk for a robot? Could it be that the robot has become part of the band of brothers?

Julie Carpenter considers these sorts of reports in the light of what she calls the Robot Accommodation Dilemma. It is something EOD soldiers in particular are prone to. Bomb disposal soldiers have a relationship with robots that is unlike other military or civilian human–robot relationships. The dilemma is that the robot is a tool, and an expendable one – but it exists to save human life, and works and travels with the team. Inevitably the robots are anthropomorphised; they 'go through a lot'. One soldier told Carpenter how a younger guy in the team named their robot Danielle and slept cuddled up next to 'her' in the Humvee. Carpenter's studies show that soldiers quickly gain familiarity with their robots, they learn their mechanical quirks and 'personalities', and start to view the robots they use as extensions of their own bodies. EOD soldiers she interviewed said they viewed the robot as their avatar – the soldiers place themselves into the robot's body, and physically associate themselves with the robot. They are emotional when a robot is lost in action. Carpenter quotes an interview with an EOD soldier named as Jed, describing his feelings after a robot team member was blown up. 'You know, here's a robot that's given its life to save you, so it's a little melancholy,' Jed says. Perhaps it's not so strange that soldiers would attempt to rescue a trapped robot they'd grown close to. It's another example, Lehmann points out, of the path to bravery being co-opted in an evolutionarily novel situation.

The team bonding and intense camaraderie in the army facilitates bravery, but you probably have to be a risk-taker to apply for the military in the first place. So what of examples of individual, one-off bravery – how do we explain that?

*

In the spring of 2012, an unexpected heatwave baked parts of southern England.

On 26 May that year, a Saturday, a Bulgarian-born British electrician named Plamen Petkov, who lived in the south-west London suburb of Sutton, headed to the coast with his girlfriend to make the most of the weather. They chose West Wittering beach, a long stretch of fine sand and clean waters near Chichester, renowned for both its surfing and its beauty. With the temperatures hitting 28 degrees Celsius, hundreds of other people had had the same idea. One was San Thidar Myint, from north-west London, who had brought Darlene, her five-year-old daughter.

What happened that day simultaneously illustrates two extremes of the human response to disaster. Darlene was in the sea on an inflatable ring when a rip current dragged her far from shore. The girl screamed, and her mother, panicking, begged someone on the beach to help. The waves were suddenly intimidating, the currents treacherous. What would you do? You would have to balance the risk to your own self, after all. Perhaps you would tell yourself that someone else will help, or that the situation will resolve itself without you. You can imagine people shifting uncomfortably in the sand. No one on the crowded beach answered the mother's cries.

San Thidar, who couldn't swim, became even more dis-traught. This was when Plamen, aged thirty-two and happy at having recently moved in with his girlfriend, happened to walk by. Witnesses stated that he charged into the water without a second thought, despite Darlene being in a red-flagged zone with 'do not swim or enter the water' signs.

When he reached Darlene she abandoned the rubber ring and climbed on his back, and he started swimming to shore, keeping her head above the waves although the wind and current and the girl's weight meant his head kept submerging. Plamen managed to carry Darlene closer to the shore, where he passed her to a woman who took her safely to her mother. But Plamen was sucked back in and dragged under. By the time he was brought to dry land he was unconscious, and the bystanders who attempted CPR failed to revive him. He was pronounced dead. The cause of death was cardiac arrest.

The coroner's officer who dealt with the case said it was the most unselfish act she had seen.[2] Plamen was posthumously awarded the Queen's Gallantry Medal for bravery. Noting that his selfless actions saved the girl's life, the citation ends with this line: 'Once he had reached the child he did not release her to save himself even when he got into difficulty.'

Reports such as these leave us in awe of the human capacity for bravery. You hear these stories and think: what you would do in a similar situation? It might be enough to say, 'he was an incredible person, he was one in a million' – but I want to try and understand these people at the peak of their potential. Why did Plamen risk so much, ending up giving his life for a total stranger? What was different about him?

We now know a lot about what happens in the brain when people are brave, and about what happens to their hormones. We can start to explain what happens biologically. Scientists are also figuring out how to induce bravery, and how to use what we know to treat people suffering from PTSD and other kinds of stress and anxiety.

Faced with disaster or a sudden desperate situation, the brain produces corticotropin-releasing hormone (CRH). This starts a chain reaction which primes the body for action: adrenalin gets the heart pumping faster and blood sugar levels increase in anticipation of activity. CRH also interacts with the amygdala, the almond-shaped paired structure in the brain responsible for generating feelings of fear and anxiety.

For some people, fear prevents them from taking action. For others, as we've seen, fear can be overridden. It may be a case of accommodating and living with a slow-burning fear, such as that experienced by Henson and EOD operatives, or dispelling an immediate fear, as we saw with Plamen.

After he died, Plamen's friends said his act was typical of his selfless nature. This explanation is an appeal to personality, something that Bryan Strange, who runs the Laboratory for Clinical Neuroscience at the Reina Sofia Centre for Alzheimer's Research in Madrid, Spain, says is one of the two main parameters that scientists are exploring to understand why some people are individually brave even on behalf of people they don't know. The Strange Lab, as it is called, is concerned with how the memories of traumatic events are stored in the brain – and how they can possibly be erased.

The other parameter is genetic polymorphisms. 'There are genes under study, such as the *ADRA2b* gene in the adrenaline system,' Strange says, 'but much like the pathogenesis of mental illness, it is likely that a combination of genetic polymorphisms are behind differential fear responses.' It reminds me of what geneticists told me about intelligence: there are likely to be many different gene variants that influence a complex behavioural trait such as this.

Gleb Shumyatsky is a geneticist, with his lab at Rutgers University in New Jersey. His research area is the molecular and cellular analysis of fear and memory, but Fear and Loathing in New Jersey would seem to fit – Shumyatsky studies the genetics of fear. His team have identified a protein, stathmin, that plays an important role in the amygdala, linked to regions known to influence the fear response. Stathmin is produced by the *STMN1* gene, and mice bred without it explore more of a new environment, and show more 'boldness' in entering open areas. As well as this lack of what the researchers called innate fear, experiments showed that the mice were unable to form memories about fearful events. And female mice without the stathmin gene showed different maternal behaviour: they did not retrieve pups from harmful situations, and in experiments didn't bother locating safe places to hide in the face of a threat.

Fear is an adaptation – we need fear. In the wild, of course, fearless mice would not last long, but making a mouse in the lab without the stathmin gene allowed the scientists to examine how fear is generated and processed. The aim is to find treatments for people who have pathological fear or suffer from conditions such as PTSD.

Would this allow us potentially to damp down the fear response? 'One can train certain simple behaviours as a response to dangerous situations,' says Shumyatsky. We know, he tells me, that mice can change the activity of the stathmin gene. As we've heard, mice without the gene are braver, and ramping the 'volume' of the gene up or down modulates their levels of fear. Shumyatsky points out that in people who have an exaggerated response to fear, there are mutations in genes that control this volume switch. So might particularly brave people such as Plamen have variants of stathmin that make them innately more courageous than the rest of us? The problem, Shumyatsky says, is that scientists generally study people who are inordinately fearful rather than those who are super brave. 'I have not heard of stathmin gene variants found in brave people, but potentially it is possible.'

*

Some people act without apparent fear. Bomb disposal soldiers, as we've seen, learn to live with it. As the slogan has it, they feel the fear and do it anyway. Some people, such as Plamen Petkov, are able to show remarkable bravery and it's not because they don't feel scared, it's because they can override their fear. But there are some people, very few worldwide, who are biologically incapable of feeling fear. Are these people brave, as well as fearless?

Justin Feinstein is a clinical neuropsychologist at the Laureate Institute for Brain Research, University of Tulsa, Oklahoma. Feinstein works with three women famous in the neuroscience literature, but known only by their initials: SM, AM and BG. These women have Urbach-Wiethe disease, a rare genetic disorder that has various effects on the body, such as the thickening of the vocal cords, but which also causes calcium deposits to build up in the brain. In SM's case, the disease is caused by the deletion of a single letter of the DNA code in the gene coding for extracellular matrix protein 1 – a protein with a number of jobs in the body, which when it goes wrong in other ways can cause breast and thyroid cancer. For an unknown reason, the single-letter deletion in Urbach-Wiethe disease causes calcifications that seem to target and destroy the amygdala but nothing else in the brain. What happens if you lose the structure that generates fear? You lose fear itself.

Feinstein has now known SM for many years. Before the disease progressed, as a child, SM recalls being scared by a vicious Doberman that cornered her – it caused a 'gut-wrenching terror' – but in her adult life she has never experienced fear. For example, on one occasion a man she had never seen before held a gun to her head and shouted 'BAM!' Someone else witnessing the menacing scene reported it to the police but SM was perplexed when they arrived, saying only that she found it strange that someone would do that.

On another occasion, a man, probably demented on meth or crack, put a knife to SM's throat and threatened to kill her; and once she was tricked into getting into a car with a stranger who took her to a deserted farm and attempted to rape her. She shouted at him to take her home, and the appearance of a dog attracted to the noise caused the man to give up his attempt. Asked later by a shocked and worried Feinstein if she'd been scared, SM said no; she'd just been angry. Her complete lack of understanding of danger meant that she got back in the man's car and let him take her home, even directing him to her apartment so he knew her address. In many ways it's lucky her fear deficit has not caused her greater harm. 'Her impoverished experience of fear repeatedly leads her back to the very situations she should be avoiding,' says Feinstein. As it is she has survived for more than fifty years.

Feinstein and some colleagues once took SM to a famous haunted house at a theme park. She led them along dark passageways, excited and curious but showing no signs of trepidation nor feeling any fear. Other people in their group shrieked and jumped at the surprises, but SM was unmoved – she never screamed or jumped back or even flinched. 'In the haunted house I had a distinct impression that she was leading me into battle,' Feinstein says, 'although I am afraid we would not have survived very long with SM leading the way.'

Bravery without danger is showboating. Action without fear is foolhardy. But trauma without fear goes unlearned. Feinstein notes that after some traumatic event, even the assault and having a knife held to her throat, SM shows no sign of dwelling on it. She doesn't start avoiding similar situations, nor does she remember the event with any drama or emotion. It seems that without her amygdala her memories lack any fearful dimension and she fails to learn from them. Feinstein points to a study in

2008 showing that veterans of the Vietnam war who suffered brain injuries in battle – and specifically, damage to the amygdala – did not develop PTSD.[3] So in extreme examples it might be beneficial not to have an amygdala, but what neurologists and psychiatrists take from this – and the dozens of scientific papers inspired by SM over thirty years – is that if we can modulate the action of the amygdala, we may be able to develop better treatments for PTSD and other disorders.

PTSD is a big problem. According to the US National Center for PTSD, 7 to 8 per cent of the population will experience it at some point in their lives. Symptoms include persistent terror, reliving and remembering the traumatic event, problems sleeping, a feeling of detachment and a propensity to be easily startled. Women are more susceptible than men, and about 8 million adults will have it in a given year.

Eventually Feinstein hit on a way of scaring SM: he had her experience suffocation. In the safe but undoubtedly terrifying experiment, SM donned a mask that fed her air containing 35 per cent carbon dioxide – that's some 875 times the amount in the air we breathe normally. As soon as she began breathing the suffocating mixture, she started gasping. After eight seconds she was waving frantically; after fourteen seconds she exclaimed 'Help me!' The scientists removed the mask. Two minutes later SM stopped talking and started having more trouble breathing. She tapped at her throat and gasped, 'I can't.' She was having a panic attack, the first in her life. Five minutes after the start of the experiment, she had recovered, and reported that she had felt genuine fear, the worst in her life. 'After many years of attempting to scare SM,' says Feinstein, 'we had finally found her kryptonite: carbon dioxide.'

Feinstein later repeated the experiment on an identical twin pair of German women, AM and BG, with severely damaged

amygdalae, and got the same result. It's odd to think how deliberately inducing a panic attack in women can be celebrated and
seen as a scientific breakthrough, but until then, the amygdala
was seen to be absolutely key to the experience of fear. How
then could these women feel fear, without the amygdala? The
answer seems to be that there are other, more primal pathways
to fear in the brain. Although the amygdala is important in
conducting fear responses, it is not essential.[4] And although
without the amygdala we don't feel fear, being fearless is not
the same as being brave.

Before his work with the Urbach-Wiethe patients, Feinstein
treated US military veterans who were suffering from PTSD.
'I did psychotherapy to get them to overcome some of the fears
they'd acquired during war,' he says, referring to deployments
in Iraq and Afghanistan. One way to treat PTSD is to assist the
patient to remember and re-experience, to a small degree, the
stimulus that caused the trauma. Prolonged exposure therapy,
as it's known, involves both discussing the memory and revis-
iting it in a controlled and safe way. For example, Feinstein
recalls veterans with PTSD who had been in vehicles blown up
by IEDs in Iraq. Typically sufferers are unable to drive or even
to go near cars, because vehicles trigger anxiety and panic.
And if that's not enough of a life-ruining, miserable disorder,
it means you can't have a car. 'If you can't drive in America,
you're screwed,' Feinstein says. 'You might start treatment by
having the veterans look at cars on computer monitors. That
was fine, but they freak out in parking lots. Or they are okay
as passengers but freak out as drivers.' So you build up slowly
and safely, over months. 'It's a fascinating treatment to deliver
because essentially what you're trying to teach them is how to
overcome their fear. It's done systematically, it's like a class, and

you have to have intuition about their limits. You have to make sure when you induce the fear response it's in a safe context. And continue inducing it until it's no longer there.'

The opposite works too – you can prepare the brain to guard against fear. It's why training works. I'm recalling stories of cabin crew evacuating crashed aircraft, and firemen rescuing people from burning buildings. 'You can damp down fear through training,' says Feinstein. But instead of the cabin crew airline training I was imagining, he tells me about the extraordinary training undergone by elite military units. At the Little Creek naval base in Virginia Beach, Virginia, for example, candidates hoping to become Navy SEALs are put through one of the most demanding and dangerous training courses imaginable. The Basic Underwater Demolition/SEAL (BUD/S) component of the course lasts twenty-four weeks. One notorious activity is known as 'drown proofing', whereby candidates must swim 100 metres with their arms and legs bound. Another exercise simulates drowning. The drop-out rate is huge, up to 90 per cent, and the training is so demanding and dangerous that, tragically, deaths sometimes occur. 'All these kinds of training are forms of exposure therapy,' Feinstein says. 'With repetition you habituate the fear response.' With repetition you can even get used to the horror of near-drowning.

After intense training, actions and operations can be carried out 'unconsciously'. Hence the phenomenon many of us have experienced, when we drive home by a familiar route and find that we have no memory of the journey. The operation is being conducted not by the cortex, where most of our decision-making and executive function resides, but by the basal ganglia at the base of the forebrain. This region is basically our autopilot, and it isn't influenced by the fear-mongering of the amygdala.

That's not to say dragging children out of a burning house is like driving home on a quiet road. Even with training, these feats are impressive. But the point is that bravery can be trained. Courage can be practised, like a discipline.

Bryan Strange explains how training apparently makes people braver. He says that fear responses can be dampened down consciously, by processes such as cognitive regulation or active coping. In Frank Herbert's sci-fi novel *Dune*, the main character is put through a terrifying and painful test. He copes with it by repeating 'the Fear Litany': 'I must not fear. Fear is the mind-killer. I will face my fear. I will permit it to pass over me and through me.' I remember a couple of times after reading this as a child repeating these lines when I was woken by a nightmare or faced some fearful situation.

My recollection is put in perspective when I mention this to Dave Henson. When he was blown up, he says, he automatically started listing the process that needed to be followed in the event of a casualty scenario. 'The training definitely kicked in and I think providing a little direction, even if it wasn't needed, perhaps provided a level of reassurance for those that were dealing with the immediate situation.' It distracted his mind from the pain, too. The kinds of injuries he suffered more often than not result in death, he says, and it is drilled into soldiers from all their medical training that time is of the essence. Possibly, he reckons, by demonstrating that he was still in control and thinking clearly – even if the instructions he'd been given were not necessary – he enabled the actions required to save him to be carried out in a more calm and collected fashion. The army version of the Fear Litany works. 'I did feel strangely calm,' says Henson, 'once the initial shock had died down.'

According to Strange, trained habits and practised behaviours are more easily accessed by the brain during terrifying situations: 'There is accumulating evidence that fear – or stress in general – favours the retrieval of habit memory at the expense of flexible hippocampus-dependent memories.' That is, the brain prefers to go into autopilot at times of fear and stress.

Training can change the brain and make us braver. And there's another way this can happen, but it's available only to women, and you have to change your body too: through pregnancy.

Can you imagine how you would react if, out of the blue, a masked man pointed a gun in your face?

In January 2016, Angie Padron stopped with her children, aged one and seven, at a Tom Thumb gas station in Hialeah, Florida. Padron, then twenty-one, parked in such a way that the pump was opposite the driver's side, so she walked around the car to fill the tank. At this point, as can be seen on film from the gas station security cameras, a masked man approached and levelled a gun at her head. Another masked man ran to the driver's side and opened the door. The men were carjackers.

'Instincts kicked in,' Padron told *Good Morning America*.[5] 'I was yelling at him, "My kids are in the car. Don't get in the car," and he got in anyway.' Her son, Evan, was on the back seat. 'My mom was yelling at him, saying, "Get out of the car, get out of the car, get out of the car, right now",' he says.

When her screams proved ineffective, she opened the door on the driver's side and leapt on the man trying to steal her car. In a brief struggle she ripped his mask off and managed to drag him out of the car. Both men fled. (They were soon caught by police.)

Her story made national TV and got international coverage, and it's obvious why: here's an ordinary young mother performing an act of remarkable bravery. We see Padron doing something that we hope we would also have the capacity for, if we were ever tested. We like to be reminded that there is valour potentially available to us all, even as we go about our regular lives. A hidden superhero within. So she was rightly feted for her bravery. But just as with Plamen Petkov and Dave Henson, I want to know what's going on inside.

Oliver Bosch, a neurobiologist at the University of Regensburg in Germany, studies maternal instinct in mammals. He explains what we know about how being a mother can influence behaviour. First, he says, it's well known among biologists that while lactating, rodents as well as human mothers are less anxious. (When talking about rodents, scientists prefer to say the mothers are less anxious rather than more brave, and if you're wondering how scientists assess anxiety in mice, they measure things such as the animal's tendency to venture into open or brightly lit spaces.) So mothers are less anxious, our proxy for bravery, and this calming effect comes about through the action of the hormone oxytocin. You might have heard that oxytocin has a role in promoting mother–infant bonding, which has given it the nickname the cuddle chemical, but there's far more to it than that.

Before birth, oxytocin has an important role in triggering the contraction of the smooth muscles of the uterus and inducing labour, and when the mother is nursing, the sucking stimulates the release of oxytocin which causes the milk ejection reflex from the nipple. 'But what also happens is that oxytocin gets released within the brain and facilitates maternal care and maternal defensive behaviour,' Bosch says.

His work in rodents has shown that oxytocin is released in the amygdala when the mother is in danger. 'There is a strong release of oxytocin when the mother is facing a threat, such as a potentially dangerous intruder rat that might kill the offspring.'

Oxytocin in the amygdala of men makes them brave, and Bosch speculates that this same stimulus in lactating females is responsible for maternal bravery. Oxytocin also changes the way mothers react to stress. When Padron said 'instincts kicked in', what was happening was that the encounter with danger and a threat to her children caused her brain to release oxytocin in the amygdala. This then blocked the production of the corticotropin-releasing hormone. This hormonal override gave her the courage to confront her assailants.

When a rodent's pups are weaned, the mother's behaviour returns to normal and her maternal responses disappear. That's not the case in humans. Rats can have sixteen pups every eight weeks, whereas women are currently having fewer than two babies in a lifetime. Plus human babies need far more care for longer.

'The maternal feelings of the human mother need to be present for a much longer time in order to ensure the proper development of the kids,' says Bosch. It makes a big difference between us and mice, and having children could even structurally alter your brain. 'I'm not sure if the maternal feelings ever go away once a woman has had a baby.'

In this exploration of bravery, I've realised that it comes in many forms. There's what we might call learned bravery (such as that demonstrated by Navy SEALs and cabin crew); there's maternal and familial bravery (protecting a child); there's social bravery (public speaking); and there's altruistic bravery (saving a stranger from drowning). As we've seen, there are different ways these kinds of bravery can evolve. But that doesn't mean that they are handled by different parts of the brain. At root, all forms of bravery can be seen as having a common denominator:

the voluntary performance of something that is opposed by a current and ongoing fear. So courage in different forms may share the same core brain mechanisms.

Uri Nili and colleagues at the Weizmann Institute of Science in Rehovot, Israel, set up an experiment to examine this. The team recruited people who were scared of snakes, and scanned their brains while at the same time giving the volunteers the choice to bring a live snake closer to their heads. The snakes were positioned on a conveyor belt that was under the control of the person in the brain scanner; the conveyor belt could be activated to move backwards, away from the volunteer's head, or forwards, towards it. By setting up the choice like this, the scientists were able to define a particular act of bravery: the decision to move the snake closer. 'The way we define courage, that is, "acting in a way opposed to that dictated by ongoing fear", suggests that people that are braver than others have a higher ability to control their fear,' says Nili.

He suggests too that multiple acts of moving the snake closer show perseverance in the face of adversity, which is another component of bravery. As a control, the team used the same conveyor-belt-and-snake set-up to scan the brains of people not scared of snakes. They also performed the scans with only a teddy bear on the conveyor belt.

What they found (and published in the journal *Neuron*)[6] was that the decision to overcome fear was mainly associated with activity in an area of the brain called the subgenual anterior cingulate cortex, or sgACC. The cingulate cortex is a mid-level layer of the brain, and we're interested in the anterior, front half of it. This region connects to many other parts of the brain, including the amygdala and the hypothalamus, and has roles in pain perception, in the assignation of emotions to internal and external causes, and in guiding social behaviour. The sgACC is the front lip of the front half of this part of the brain. The results from the snakes-on-a-scanner experiment point, Nili's team say, to the sgACC being the region that has to work in order to mount a successful mental effort to overcome fear. When you are marshalling yourself to any kind of bravery, you are damping down the output of the amygdala and the hypo-thalamus, and for that it seems you have the sgACC to thank. Brave people have high sgACC activity, and they are better able to inhibit the output of the amygdala and the hypothalamus during times of fear.

There's a key point I take away from this which doesn't require us to remember the names of parts of the brain. Nili's work suggests that although there are several varieties of brav-ery, there is a common denominator in the brain when it comes to mastering fear. 'Any instance of overcoming an acute source of fear, of any type, would be regarded as an act of courage, and is probably mediated by the brain processes described in the paper,' he says.

Looking back on his career in explosive ordnance disposal, Dave Henson jokes that it's a stupid job to go and do. Now, he says, he's a proper pansy. But I don't believe him. Bravery isn't just about military valour or one-off reactions to armed robbers. That's when we see extreme manifestations, but in a smaller way we use it every day. It's instructive that the rodent biologists use anxiety as a proxy for bravery. The two are linked in the brain. And it's reassuring that bravery can be trained. We might not be able to display the courage of war heroes, or the extraordinary selflessness of those who dive into dangerous waters to rescue strangers, but we can ratchet ourselves along.

Rehabilitation for Henson was long and painful. 'It took me a long time to realise it was hard,' he says. 'There were tears in hospital because it was hard.' But he got through it in the same way he got through day-to-day deployment in Afghanistan. He was in hospital with a lance corporal and two privates from the Parachute Regiment, and they did it together. The team had supported him through the deployment and the accident, and now an impromptu hospital team supported him through rehab. 'In that situation one of the lads might be having a really shitty day and I'll help them out, then I might be having a shit day and he'll help me out. And that's how you get through it. It was one of life's pleasures to be in hospital.'

After Henson was blown up he was promoted to captain and then obviously had a desk job. At least he could use a wheelchair from the off, the result of his eerily prophetic decision to learn to use one when writing his dissertation on disabled people taking up sport. It's mad that for his engineering degree he wrote about amputees and sport. 'It's barking mad,' he says. Team sports played a massive role, and when he was fitted with running blades he became even more motivated. In the army there are twice-yearly fitness assessments, for which you must be able to run 2.4 kilometres in under 10 minutes 30 seconds. Henson decided he would leave the army once he was able to pass this test, and in 2014 he ran 10:28. Two years later he was in the Paralympics in Rio de Janeiro, and he won bronze in the 200 metres. He ran it in the extraordinary time of 24.74 seconds.

By his own reckoning, Henson is lucky. But he's not referring to his survival after being blown up. It's the re-establishment of his character and the redefining of his purpose. 'Everyone cares what people think about you,' he says. 'It doesn't matter what they say, they care. For me there was that pride in being an army captain. I knew that people understood that role. But after I left the military I became a wounded veteran, and I didn't have that definition any more. And then through the Invictus games and the Paralympics – I wasn't just a wounded veteran, I was a former serviceman and a medal-winning Paralympian.' He is lucky he had something he could do. As well as sport, he's currently doing a PhD in biomechanical engineering at Imperial College London. He's researching how to build bionic knee joints for amputees, and hoping to use that little bit of his left knee he still has in a new kind of prosthetic. 'Suddenly that redefinition is almost complete,' he says. 'The proper redefinition will happen here at Imperial. I'm starting to make a difference in people's lives again.'

Have we covered all forms of bravery? I feel I am much clearer now about what bravery is, how it evolved, and what the body needs for it to happen, physiologically and neurologically. My fuzziness over bravery has decreased, but my awe at its power hasn't. If anything, thinking harder about what makes people brave has only increased my admiration. About random acts of bravery, Feinstein says he's always been struck by how quickly people can act without conscious thought, seemingly without deliberation. 'It seems primal and instinctive. It might be more ubiquitous than we imagine.' We might all have it inside us. Now that's a nice thought.

6

SINGING

Time is a strange thing
You live your life and don't notice it
Then suddenly it's all you can feel.

Richard Strauss, *Der Rosenkavalier*, 1910

Sweet as my own milk
But salt as my own tears.

George Benjamin and Martin Crimp,
Written on Skin, 2012

I'm at the Royal Opera House in Covent Garden watching *Der Rosenkavalier*. If, like me until recently, you've not heard of it, it's a comic but melancholic opera written in 1910 by Richard Strauss. This evening's performance started at six, and it's now nine thirty; we're in the third act. My eyes are starting to blur. Don't get me wrong — the performance is wonderful, I've got an amazing seat in the stalls, the orchestra is as stirring as you'd expect from this world-class group of musicians — but it's a long opera. I'm tired. Then I guiltily reprimand myself. *I'm* tired? What must it be like to sing it?

A few hours earlier I popped backstage to say hi to the bass operatic singer Matthew Rose, who is playing Baron Ochs. He was in make-up, having a latex bald head fitted and coloured to match his skin. On top of it went a wig, and on top of that, a hat. You can imagine your head would get rather warm, and that's before going on stage. Ochs is a swaggering, cynical and crude member of the Vienna elite; he's a central figure in the opera and sings through all three acts. Rose told me the role is a monster, and he didn't mean the man's character. 'There are a few Wagner roles that are as demanding but this is probably as hard as anything I'll do. This is the Everest of opera. This is the ultimate.'

I'm far from being an opera expert, but I know I like it, and I know why. In addition to the straightforward enjoyment of getting wrapped up in a great story, it's the thrill and the privilege of seeing people performing at the peak of human ability. It took me a long time to realise that, but it's why seeing the opera (or the ballet) in real life is so moving: you're experiencing the achievement up close, almost sharing it; you're with someone who has learned to use their entire body as a musical instrument. As a biologist, and as someone who can't sing, that is powerfully impressive. But until meeting Rose, I hadn't appreciated just how *difficult* it is to perform opera. You have to have absolute control over the production of sound. That means your breathing, how the shape of the larynx affects vibrato and the sound of words and vowels, how your chest voice (the lower end of your register) merges and interacts with your head voice (the upper end). But that's not all.

'You have to be able to perform in different languages, you have to be able to learn complicated music in a foreign language, you have to be musical with it, you have to watch a conductor, you have to be on stage with other people,' Rose says, and I'm starting to feel exhausted just thinking about it. 'It's amazing the number of things you have to be able to do at once at an unbelievably high level. Simon Rattle says being an opera singer is the hardest job in the world. Singers get a hard time but what they have to do, it's the hardest skill set of any job anywhere.'

Their job is to faithfully re-create someone else's music. It might be Verdi or Wagner, but someone else has determined what they want to do with the opera, and the singer has to transmit that, not interpret it. In pop music the singer is letting his or her emotions come through and it's okay to lose control – that's what moves us, when we hear the emotions breaking through in the voice. But in opera, the emotions come from the composer; control is absolute. Plus you have to sing without a microphone. As much as I love pop singers, they simply couldn't make themselves heard in an opera venue over the sound of a full orchestra without a mic. Opera is full on. Technically, it is more complex than acting, because singing is more complex than speaking. And in theatre the actor is in control of time, and spontaneity can be played with, cues can be adjusted. 'In opera there isn't that elasticity of time. You have rehearsed it all with the conductor so spontaneity in opera is largely faked,' says John Fulljames, director of opera at the Royal Danish Theatre.

I suppose that because opera is something I could never possibly come close to performing in, those who do so carry some sort of aura, an almost magical talent. It's the same with anyone with an extraordinary voice. She has a marvellous gift, we say. He has god-given talent, we say. To put it scientifically, we assume there is a strong genetic component to excellence at singing. But here I am with Matthew Rose, a Grammy Award-winning singer – let's ask him.

'I really believe that 90 per cent of what I'm able to do is because of my training. Language skills, musical skills – I don't think that can be genetic.'

Well, that told me. Of course, Rose and all professional musicians have worked incredibly hard to get where they are. And despite the blithe 'he has an amazing gift' way of describing people with talent, it's also popular to ascribe expertise to practice, as the infamous 10,000 hours theory (see below) holds. Can both explanations be right? My aim now is to explore what we know about how people reach the peaks of singing ability. What does it take to become an opera singer of the calibre that gets gigs at Covent Garden? I'll speak to singers and teachers but also to geneticists, because when talking to Rose I felt that his insistence on training over genetics comes from an understandable need to acknowledge the amount of hard work he's put in, especially for the role he's currently working on, but is also somehow his way of expressing modesty in his talent. Anyone could do it, he virtually says. Well, I couldn't.

Rose says he was not born into a particularly musical family. 'My mum actually has a beautiful singing voice, and retrospectively now she loves music,' he says. Music was only 'around' at home and he was at a school where music only 'existed'. But he did enter the school choir from the age of seven. 'That was a major influence in what I do and there was a lovely music teacher at school who was another major influence. So I was always singing but I never really took it seriously all the way through my teenage years.'

What he did take seriously was first swimming, then golf. His dream was to be a professional golfer, and I think it's instructive that he switched, after a while, to singing. We'll see later how important it is that children try different things so they can get a feel for what they like *and* what they are good at. 'Thank God I didn't do golf,' he says, 'I'd be selling Mars bars or something by now.'

When he was around seventeen a new music teacher at school suggested Rose should think about singing as a job. Rose calls it a job, not a career, much less a 'calling'. 'Obviously I had a natural voice, and some musicality and when I was at university I was made aware of what it could be.' He then 'luckily' got a place at the Curtis Institute of Music in Philadelphia, and he attributes much of his subsequent success as a singer to the training he received there. 'Basically I started from scratch when I went there and had five years of unbelievable training. It was an amazing environment to learn the tools necessary for this trade.'

He says he didn't know much about opera. 'I went there, I could sing three or four notes, I had no clue about anything.'

The competition for places is intense, and the calibre of the students is extraordinarily high, so I'm not sure how much luck was involved in him getting into the world-renowned music school. Lang Lang, one of the world's most famous pianists, was in the same year. At Curtis the training is vocational, and perhaps this is why Rose views what he does as merely a job. Perhaps it's just that I see singers as having this aura of talent – or maybe I was half-expecting a stereotypical diva, occupying some higher plane of artistic existence. Anyway, at Curtis the singers performed five operas a year, so classes were always about training towards the performances. 'It was completely vocational training,' says Rose, 'which is unprecedented in the world. It's like being a footballer – you've got to get out there and do it, sitting around learning about it in a classroom doesn't really help.'

If this isn't concerted, targeted, deliberate practice, I don't know what is. It gets to the heart of the debate about expertise, so let's look at the evidence, some of which we touched upon in Chapter 1.

On one side there are the environmentalists. These are the people who argue that practice, and particularly what is called deliberate practice – focused, dedicated practice with targets and improvement goals – is the most important factor in the development of expertise. They are led by Anders Ericsson, who says: 'No matter what role genetic endowment may play in the achievements of "gifted" people, the main gift that these people have is the same one we all have – the adaptability of the human brain and body, which they have taken advantage of more than the rest of us.'[1]

The writer Malcolm Gladwell took a 1993 paper by Ericsson and colleagues [2] and from it developed the '10,000 hours rule', the idea that it takes that amount of time to become an expert in something. Ericsson took issue with simplifying his research into a 'rule',[3] and has complained that Gladwell didn't even mention that it is *deliberate* practice that is important, not just any old practice.[4] However, the meme was too good, and it took off.

On the other side are those who give space to innate talent – who allow that some people are genetically gifted. They look at the evidence and see that expertise is best explained by a mixture of genetic and environmental factors. Their framework for expertise – and let's face it, it's not going to compete well on the meme front with 10,000 hours – is called the multifactorial gene-environment interaction model, or MGIM.[5] In a nutshell it recognises that practice can't explain achievement, and accepts that both genetic and non-genetic factors are essential for expertise. The model has been developed by Fredrik Ullén and Miriam Mosing at the Karolinska Institute in Stockholm, Sweden, and Zach Hambrick, who we met in Chapter 1. Hambrick runs the Expertise Lab, devoted to uncovering the reasons for individual differences in expertise. Why, his team ask, are some people so much better than others at certain things? 'We look at training, experience and talent – this is basically the capacity to be influenced by genetics – and also socio-demographic factors,' he says.

'No one doubts that you have to practise in order to become an expert in something like chess or athletics or music. We aren't literally born with this knowledge. But what accounts for differences across people? That's what we focus on.'

They've amassed a lot of empirical evidence that they say supports the idea that practice is not enough to explain performance. So let's look at a couple of papers. In one, Hambrick's team took eight separate studies of musical ability and practice, and analysing data from all of them, found that the amount of practice someone put in accounted for 30 per cent of the variance in performance. In other words, factors other than the amount of practice put in accounted for 70 per cent of the individual differences in performance ability.[6] In another, more wide-ranging, analysis of musical ability (not carried out by Hambrick), practice was found to account to for only 36 per cent of the variance in performance.[7] Many other similar analyses have led the MGIMers to the view that genetics is just as important as – if not more than – practice. So despite Rose's assertion that 90 per cent of his success is down to training, the evidence shows that on average only 30-odd per cent of the variance in performance outcomes can be explained by practice.

Rose gestures to his throat. 'These are the most precious muscles in the body.' He then gives me a quick primer on how an opera singer produces such volume. You know how babies are able to make an unbelievably loud noise? That's because they use their entire bodies efficiently to produce sound. 'We're trying to regain that efficiency, as a baby would,' Rose says. 'It's all about using your diaphragm and the right muscles and pushing the sound in the right places.'

Having the right vowel sound is an advantage, because it gets you close to the resonances you need to produce to fill an opera hall. 'There's a reason Italians and Welsh are such great singers,' he says. 'It's because the way they speak is already very close to how you have to be as an operatic singer trying to produce those resonances.' Being French, for example, is no advantage (the language is guttural, apparently), and an accent from the south of England, where Rose hails from, isn't much better. The short vowel sounds of the north of England are better (which is why the composer Alan Williams, from Salford University, has created an opera designed for the northern accent).[8]

The role of Baron Ochs is long, and sung low in Rose's range, which means his voice needs to be completely rested before a performance: he won't even speak in the day before he sings.

'I'll wake up late, have a nap in the afternoon then come in. It's just so tiring. It's physically, mentally demanding. This is the marathon of opera roles. So I have to be so careful. It's unlike anything else I have to do.'

It's an endurance test, and he has to be conservative just to get through the night. When he was young, adrenalin would see him through, but these days he sees it more as a job he has to turn up for and do to the best of his ability.

This sounds dour, but it's not that he doesn't love it, it's simply that he can't afford to waste the energy on being excited. In New York in 2016–17 he played Leporello in Mozart's *Don*

Giovanni. 'Doing Don Giovanni, I can do that falling off a log so I got really excited. It's less complicated and I know it really well. But with this you have to concentrate every single second and watch the conductor. The text is in German but I have to do a slight Austrian accent. So this role is as hard as it gets.'

Rose again and again emphasises that, for him, it is a process. It's about learning to use his body properly, to breathe, to look at a conductor properly, all things that improve with experience. Given all that, however, he says that there *are* 'natural' singers: he names the Welsh bass-baritone Bryn Terfel as one, and the French tenor Roberto Alagna as another. Terfel started singing in Welsh-language Eisteddfod competitions when he was four, so he has had a huge amount of practice, which I would have thought makes it difficult to assert that he's a natural because we're back to the problem of practice and natural talent. Alagna started off busking in the streets before he was discovered, and despite huge success in opera, was apparently self-taught. 'Alagna never had a singing lesson,' Rose says. 'But most people have had years and years of lessons. Pavarotti, probably the most perfect singer that ever lived, he studied and studied, and got his voice the way he did though study and practice.'

Mikael Eliasen is Rose's old singing teacher. He is head of the department of vocal studies at the Curtis Institute of Music. He has worked with some of the world's greatest singers, including the late acclaimed American baritone Robert Merrill, the Israeli contralto Mira Zakai and the Swiss soprano Edith Mathis. He has been music director of the San Francisco Opera Center and artistic director of the European Centre for Opera and Vocal Art, and he works with the young-artist programmes at Curtis, the Royal Danish Opera and the Opera Studio of Amsterdam. He is emphatic on the question of innate talent:

'Some people are born with a voice that other people want to listen to. I don't think you have any choice about that, either you have it or you don't.' Take a world-famous singer like Renée Fleming, for example. When we say that she has natural talent, we mean she has a beautiful voice that she was born with. She has worked very hard to achieve her success, but it started with the natural sound, says Eliasen. 'That's what we mean when we say someone is a natural singer.' Without this innate talent someone who loves singing might pursue it as a hobby and get great satisfaction out of it, he says. 'But if they don't have that gift then probably it won't turn into a professional career.'

For the audience there can be a lot of emotion in opera but for Rose, again, it's not something he feels when he is performing. I hadn't appreciated, given how much the audience can be moved emotionally and even transported to another level, in the way that extraordinary art can give you a rush, a glimpse of the higher potential of human beings, that the performers themselves don't necessarily feel they are operating in a higher realm. They are, apparently, just doing their job.

'I don't get emotional when I'm singing,' he says. Or only occasionally. Schubert makes him a mess. 'But when I do opera, emotion doesn't really come into it. With Placido Domingo there must be so much emotion going through his body all the time. And his performance comes from an emotional place. With me I know how to push the emotional buttons of the audience but I don't feel it myself. I know what I have to do to make the right reaction in the audience.'

It sounds like a bit of a slog.

'It's my job and I love to do my job well. And this is the ultimate challenge in my job. To come here, this is my home opera house, the shrine of opera in this country, so to come here and do it well . . . I want to come here and do my job as well as I can.'

His secret, perhaps, if he won't acknowledge much in the way of innate talent, is a drive and an obsession that's always been there. 'I try to be successful and good at everything I do. Maybe that's an attitude thing that's ingrained in people that succeed. It's obsession. When I was a student it was complete obsession. Maybe now less so.'

To do this, he says, you have to have the right attitude. 'It's the same dedication as being a nun or a monk, it's that sacrificial to do it properly. It really is that hard.'

Rose has hit upon one of the most compelling pieces of evidence pointing at a genetic component to musical success.

Miriam Mosing is a professor in the department of neuroscience at the Karolinska Institute. She used a massive database of twins born between 1959 and 1985 to tease out the relative importance to musical ability of practice and genes. The popular idea is that if you take children and give them a little training, say in singing or piano, then the differences in ability between them at first will be innate. Ericsson would say that after months and years of practice these genetic differences will have been swamped by the new skills the children have learned, so that practice becomes the determining factor. Mosing says this is a radical environmentalist point of view, and that a large body of evidence suggests that practising isn't enough to achieve expertise.

Mosing and her team contacted people in the twins database and asked if they played an instrument or regularly sang, and if so, for how long they'd practised at different ages. The team ended up with responses from 1211 pairs of identical twins and 1358 pairs of non-identical twins. The participants were also scored for musical aptitude: pitch, melody and rhythm discrimination. Because twins share either half or all their genes, and almost the same environment, it's possible to examine a trait — such as musical ability — and determine how much is influenced by genetics and how much by practice.

Mosing found that between 40 and 70 per cent of the difference in music practice was genetic. Music practice, mind you. They found, in other words, that much of the *propensity* to practise can be explained genetically. But even more surprising for the common-sense notion that practice makes perfect, Mosing found that in identical twins, different amounts of practice did not contribute to different levels of musical ability. It's like we saw in the chapter on intelligence: if you raise identical twins separately from birth, their educational attainment is correlated more closely to each other than it is to their school or home environment. She published her findings in the journal *Psychological Science*, and titled the paper, 'Practice does not make perfect: no causal effect of music practice on music ability'.[9]

Mosing's work followed another study from Hambrick's Expertise Lab of 850 pairs of twins, who in the 1960s had been surveyed and questioned on the amount of musical practice they put in, and the success they had achieved. Remember that the common example cited to support the 10,000 hours rule is the time racked up during The Beatles' intense period of playing together in Hamburg.[10]

Hambrick and Elliot Tucker-Drob, of the University of Texas, Austin, looked at data collected for the National Merit Scholarship test of 1962. Each twin in the survey, with an average age of seventeen, had provided data on the amount of musical practice they undertook, and on the level of success they had achieved. The categories ranged from 'received a rating of good or excellent in a school/county/national contest' to 'performed with a professional orchestra'. With their data from both identical and non-identical twins, Hambrick and Tucker-Drob were able to tease apart the effects of genes and environment on both music practice and accomplishment.

They found that about a quarter of the variation in the amount of practice people put in could be explained by genetic factors. That means that a quarter of the drive to go and practise is genetically influenced. They also found that practice magnified the effects of innate talent. In short: genes influence how much you practise, and also how successful you end up being.[11]

I've never had a singing lesson, and never sing except in the car, alone. My partner marvels that I love listening to music but have no aptitude for it whatsoever. When I lived in Japan, however, I regularly went to karaoke. I'd go out with my colleagues after lab meals or scientific conferences, and I was often asked to sing 'Yesterday'. We want to hear it sung by a native Brit, my Japanese friends would urge. Poor them, to hear what is still a massively popular song in Japan, mangled by me. But over the months I found something strange happening. My voice was improving. I learned to choose songs that fitted my limited range – my go-to songs were 'Starman' and 'Light My Fire' – but also I was just getting better. Practice was helping even me, although we are talking only marginal improvement here. Even in a karaoke bar, the contrast between the noise I made and the stirring sound coming from someone who could really sing was extraordinary. At the end of their paper, however, Hambrick and Tucker-Drob conclude with a line that makes me wonder if, in an alternate-reality timeline, even I could have been a pop star: 'These results indicate that children who do not engage in training or practice in music may have hidden talents, or at the very least potentials for talent, that go unrecognised and unrealised.'

Mosing solves the riddle of why so many people, including professional singers such as Rose, still believe in the power of practice. 'They're right,' she says. 'Without practice they wouldn't be where they are, they have to practise a lot.'

No one is denying the vital importance of practice. But it's not enough to ensure you get to the very top. 'It's a really nice idea that anyone can achieve anything if only they practise enough,' she says. 'This idea that practice is an environmental factor we can use to overcome any limitations – it's an extremely popular idea.'

Mosing guesses that many of the people I'm meeting for this book probably have amazing innate talent that they start nurturing from an early age – or their parents start nurturing it. 'We really like this feeling of agency. If we are successful, we take the credit, we take agency for that. We think that it's something we can take ownership for. But in many cases that's not true. The main predictor of success is the socioeconomic status of the parent.'

'The discovery of the musicality and the singing was very natural. My mother sang with us when we were kids. I wasn't a prodigy at all, but she has recordings of us singing, and we were able to recall melodies before we could recall words. We could sing before we could speak. Me and my brother and sister, we're all musicians – and my parents aren't, by the way.'

The Canadian soprano Barbara Hannigan is one of the most celebrated of contemporary opera singers. A conductor as well as a singer, she has worked with many of the world's top orchestras and conductors, and sung on the world's biggest stages.

When she was about ten, she says, she decided she was going to be a musician. It seems that, unlike Matthew Rose and his swimming and golf, Hannigan didn't have to shop around until she found something she both loved and was good at.

Later I'll see her play Agnès, in George Benjamin's *Written in Skin*. It's an extraordinary role, created for and by her, in which she plays an illiterate, downtrodden wife who liberates herself by having an affair with a younger man – and ends it by eating her lover's heart. It's been called 'one of the operatic master-pieces of our time';[12] according to *Le Monde* it's 'the best opera written for twenty years'.[13] Hannigan, said *Opera News*, is 'one of today's most astonishing musical artists'. 'Her whole being breathes music,' said Dutch newspaper *De Telegraaf*.

We meet mid-morning; Hannigan is singing this evening. She has a light scarf around her neck, and I think of when I've seen Aretha Franklin swaddling her neck before a performance, and what Rose said about the importance of the vocal cords. Preparation for a performance is gradual.

'I warm up very slowly, with slow stretching, precision breathing, vocalising, humming humming humming.'

She hums a scale at this point and the hairs on the back of my neck stand up. (This feeling, by the way, the delicious shiver down your spine that you can also get if someone whispers in your ear, is called the autonomous sensory meridian response – and is actively sought out by some people.)[14] Quite mesmerised simply by her humming, I miss the next few things she says, so it's lucky I'm recording our conversation. 'The warm-up is about awakening the total instrument. It's not about my rib cage or my vocal cords or my breathing, it's about everything, my entire emotional being, sensual being, intellectual being, physical being – so you wake everything up, you stimulate everything in one go.'

When I hear her sing I remember Matthew Rose's description of a baby using its entire body to produce a loud noise – not that Hannigan sounds like a baby wailing, but that despite being small and slender, she produces a sound out of proportion to her frame. She has incredible control over her voice – she calls it 'the instrument', and it occurs to me that it's like playing a synthesiser, or sitting at a mixing desk, such is the control she has over it.

There's something important that this description doesn't capture. It's the magic of the performance, to use a word that couldn't be much less scientific but conveys what I mean. She has a charisma and authority that is captivating – spellbinding. I want to find out where this comes from without losing all objectivity, so I ask how she went from being a regular singer to a professional soloist.

'I always had the passion for the music. Music for me is like food: I would never eat food I don't like – unless I was starving. I only agree to sing music that I like. It's all done from the passion, and the passion gives the energy and the possibility, and it gives the energy to do a backflip, so you're so excited you can do the crazy stuff. And if you're bored – like doing my taxes, I'm bored – you can't do the crazy stuff. But if I'm passionate about something I have extra strength.'

Her parents always said she had an inordinate amount of joy, that she brought extra joy. That was the word they associated with her. Maybe it's something like this that makes the difference, that sets her above a regular, trained, very good singer.

'I think so. Because I go into work and I think, my God, what an incredible place to be, to walk into work with these incredible musicians and do what I do.'

What does she think comes naturally to her?

'We worked hard and I studied piano and I sang in choirs

but the ability – what I was born with was a certain energy and drive to develop the discipline to do what I do. And curiosity.'

Like Rose, she emphasises how hard she's worked. 'Sometimes if a conductor says to me "you're a genius", I think no, because I've worked exceptionally hard. The desire to work hard comes from the love for the work.'

I ask Miriam Mosing what her work tells us about the factors that influence musical success. 'One is our genetic endowment,' she says. 'And that includes this ability – drive – the ability to work really hard on one thing.'

Both Mosing and Hannigan have converged on what Angela Duckworth, professor of psychology at the University of Pennsylvania, Philadelphia, calls grit.[15] This is the amount someone strives to achieve something – their perseverance – and the amount they love what they do – their passion.[16]

Mosing points out something that is obvious the moment you say it, but is ignored whenever we talk about nature and nurture, genetics and environment. 'I think nurture really misses something here. Our nurturing environment is highly correlated with our genes. We get genes from our parents and our parents have created an environment which is influenced by their genes. In a sense the environment we arc born into correlates with genetics. If I have high musical skills, my parents have high musical skills.'

Here's an example of how Hannigan's childhood environment seems to closely correlate with the drive she feels she was born with. Her mother had three children in fourteen months ('Can you imagine? Three children in diapers?') and had to be super organised. 'She had a long schedule on the refrigerator with each of our names, and a roster: it was to the minute. Wake up, Barbara brush teeth, Ryan practise piano, everything tick tick – every moment of our day was planned, play time, bed time – who could get their fifteen minutes of piano practice before school if she didn't do that?'

The regimented scheduling has stayed with her. Hannigan's daily routine is precisely planned, weeks in advance. 'I schedule my work like that. My practice sessions – all the stuff I have to do. I had a kind of discipline from a very young age that made me want to continue that, and I did.'

As we've seen, no one denies the importance of practice, or the need for grit. But what the proponents of ultra-environmentalist views seem to miss, as Mosing's work has shown so dramatically, is that practice and grit are themselves genetically influenced. The race now is to identify the genes themselves – and it's a race that's going on around the world.

Mongolia, sandwiched between China to the south and Russia to the north, is the most sparsely populated country in the world. For the most part it is ethnically unmixed, families are large and the environmental influence – the diet, the healthcare and the type of schooling the children get – is pretty much the same for everyone. This makes it perfect territory for gene hunters.

Dashbalbar district, in Mongolia's Dornod province, has about 0.37 people per square kilometre. (By comparison, the UK has 256 per square kilometre; the US, with its much larger land mass, still has 35 people per square klick.) Despite the paucity of humans in the region, Jeong-Sun Seo at the Genomic Medicine Institute at Seoul National University in South Korea managed to recruit 1008 people from seventy-three families, and gave them a series of tests designed to investigate genetic influences on musical ability. Seo is a member of the GENDISCAN study group (the acronym stands for GENe DIScovery for Complex traits in large isolated families of Asians of the Northeast).

All the subjects had their genomes sequenced, and were tested and scored for musical skill. Individuals performed pitch-production accuracy tests, where they listened to a tone through a headset and had to reproduce it. Seo's team then performed a genome-wide association study (GWAS) on the data set. This is a test that looks for links between particular genetic variants and a certain trait. It is often used to search for genetic influences on disease, but in this case the method managed to tease out links to musical talent. For example, the team found evidence for a link between a gene called *UGT8* and known to be active in the development of the brain, and musical ability.[17] The gene is located on chromosome 4, in a region that had also been highlighted by a Finnish study investigating the genetics of musical talent.[18]

Irma Järvelä studies the genetics of musical traits at the University of Helsinki in Finland. Since we know that musicality varies between people, Järvelä's group has constructed a database consisting of ninety-eight family trees containing nearly a thousand people to look at the genetics underlying this variation. She has assessed musicality using three tests and subjects have completed extensive questionnaires, as well as providing blood samples for DNA analysis. 'We found that about 50 per cent of the inheritance of music test scores can be explained by genetic factors in Finnish multigenerational families,' she says on her website.

Järvelä has also crunched genetic data from more than a hundred studies on musical and sensory genetics (from birds and other animals as well as humans) and highlighted the genes most likely to have an effect on musical ability. They include genes involved in cognition, learning and memory and several to do with neuron function and activity.[19]

Such studies are still very preliminary, and have limited

'power', meaning the links are as yet tentative. They don't always agree, either – Järvelä's latest work turned up several genes on chromosome 4, but not *UGT8*, the gene found in the Mongolian study.

Miriam Mosing warns not to read too much into these kinds of studies yet as the samples are still too small. To find robust examples of genes linked to any complex trait is difficult – we saw this in the chapter on intelligence and we'll see it again with happiness – which is why the identification of a candidate gene in one study isn't replicated in another. She says that a few years ago geneticists thought they would be able to find genetic variants with big effects on particular traits, so there would be one variant for music skill, say. 'But the past few years have shown that for most complex traits the variants we find explain a maximum of one per cent of the variance.'

Just as with genetic variants linked to intelligence, there will be many hundreds linked to musical ability. That doesn't mean they aren't there, nor that people will stop looking. 'I doubt that in the future we will be able to test a person and say "this person should pursue a musical career because they have these specific genes",' says Mosing. We might be able to look at a person's entire genome and predict whether there are enough variants to add a little to music skill, but we won't be able to predict the outcome.

Gary McPherson is the Francis Ormond professor of music at the Melbourne Conservatorium of Music. He is also the director of that institution, which is the oldest and most prestigious devoted to music in Australia. Barbara Hannigan and Matthew Rose told me that they rated their hard work over their natural talent, but in his career McPherson's seen and taught many thousands of music students and professionals. 'Professionals in all walks of life have put in a lot of hard work to get to where they are, but for me, there are natural abilities that impact on our development, and possible genetic influences that also shape our abilities,' he says. 'I'm very interested in motivation – what drives a person to want to do the 10,000-plus hours of practice and exposure to get to that level. You can't explain high level singing or music performance ability merely through just the hours undertaken practising.'

McPherson refers me to a model of ability developed over the last three decades by Françoys Gagné, now retired but for many years a psychologist at the University of Montreal. The differentiating model of talent and giftedness (DMTG) acknowledges that the basis of ability is biological – that is, that there is a solid genetic foundation to musical skill. Here's an extreme example of what that means. It's the story of a blind boy referred to as LL, who has an IQ of only 58 but began playing the piano aged eight:

> One evening, at about age fourteen, LL heard Tchaikovsky's Piano Concerto No. 1 for the first time as a theme song to a movie on television. To his foster parents' complete astonishment, LL played that piece back flawlessly from beginning to end later that evening, having heard it just that one time. Since then LL's piano repertoire, completely from memory, has expanded to thousands of pieces. Professional musicians who have witnessed LL's piano playing have indicated that he seems to know 'the rules of music' instinctively and innately.[20]

Yi Ting Tan is a colleague of McPherson's at the Melbourne Conservatorium of Music. She has reviewed the scientific literature on the genetic components of musical ability and found several genes implicated in success in singing, musical perception, absolute pitch, music memory and listening, and even choir participation. The genes are scattered around, on chromosomes 8, 12 and 17, with a cluster on chromosome 4.[21]

'The shape of our vocal tract and length of vocal folds are genetically influenced, and I think vocal timbre definitely has a genetic basis too,' she says, pointing out that family members often sound alike. 'It is probably necessary to recognise that some individuals just have better vocal characteristics to start with, which flourishes further with training.'

Tan adds that even personality traits like motivation and conscientiousness are likely to have a genetic basis.[22] It seems you need the right genes to succeed as a professional singer.

I mentioned earlier that MGIM – the multifactorial gene-environment interaction model – is not exactly a memorable way to sum up what the wealth of science is telling us about how we become experts in a particular field. I was pondering this when I came across an amazing piece of work by Arne Güllich at the department of sport science at the University of Kaiserslautern in Germany. Güllich performed an analysis of the results of international top athletes. He compared eighty-three people who had won medals at Olympic games and international competitions (the sample included thirty-eight Olympic and world champions) with a similar number of professional athletes who were highly accomplished but had not won medals. He matched them for age, sport and gender, and recorded, using questionnaires, the amount of practice and training they'd done. It turns out that the medallists specialised *later* in their main sport than did the non-medallists, and that they had done *less* training in that main sport as children and adolescents. Now here's the thing: the medallists had done *more* practice or training in other sports than their main sport. They kept up participation in these other sports and specialised later in their main sport. Güllich published the piece in 2016 in the *Journal of Sports Sciences*.[23]

If you're wondering why I've veered into sports science in a chapter about music, Güllich says the same pattern is seen in other domains: 'The principle of "multiple sampling and functional matching" is not unique to elite sports but is also found in music, arts, and science' – although that's based on his understanding from his colleagues in those fields, not from hard data such as he has gathered for sport.

Some evidence comes from Dean Keith Simonton, professor of psychology at the University of California, Davis. Simonton looked at 911 operas composed by all fifty-nine of the classic composers, from Beethoven to Mozart, Verdi and Wagner. He found that the most successful composers were those who'd mixed up their genres. 'It may be,' wrote Simonton in his paper on this subject, published in *Developmental Review*,[24] 'that intellectual cross-training may have the advantageous function of mitigating the negative effects of overtraining.'

Güllich says the mechanisms that produce this effect are not clear, but that trying various things is effectively the same as not putting all your eggs in one basket. 'It increases the probability to settle on a main sport that particularly "fits" with the athlete. The fit might be in terms of performance, but also finding the right coach and the right peer group. It also reduces the risk of overuse injury. When an athlete develops across a range of sports, he or she can draw on experience in a variety of learning modes to develop the training regime that suits them best.'

He also has some interesting points to make about Anders Ericsson's views on expertise, pointing out limitations of the initial 1993 paper. For example, Ericsson and colleagues used junior musicians, not top-level soloists, in their study. And 'deliberate practice' was ascribed to the musicians' childhood activities, but such criteria cannot be recorded retrospectively. 'Today, numerous works have debunked the validity of Ericsson's deliberate practice theory, as far as elite sports is concerned,' says Güllich.

For his part, Ericsson has issues with Güllich's paper and with Hambrick's work.[25] The main one, he says, is that those authors don't distinguish between deliberate practice – that is, teacher-led instruction and designed individual practice – and any type of training or practice.[26] The argument is sure to rumble on, since what is at stake is, effectively, the American dream: that we can be anything we want to be. Not only that, but we can potentially become expert in anything. What I've realised in this chapter is that we can't. I couldn't have become an opera singer. Or an F1 driver. Or a chess grandmaster. I found what I could do, but it wasn't like *all* choices were open to me; those that were, I believe, were influenced by my genes. I don't feel this is something to be afraid of. On the contrary, it is empowering to understand the genetics and channel your resources to the right place.

Aniruddh Patel, a psychologist who specialises in music cognition at Tufts University in Medford, Massachusetts, says it is likely that some of the variation in singing ability has a genetic component. 'There is a swing in the pendulum back toward the idea that practice alone is not enough to explain extraordinary talent, as there are people who are roughly matched in amount of practice but greatly different in skill.'

With Güllich's work in mind, what advice would Zach Hambrick give to people who want to improve, or to maximise their children's chances of success? 'Try a lot of different things,' he says. It's back to the idea of gene–environment correlation. Our genes influence the activities that we engage in and the environments that we create for ourselves. If we try many different things, we'll find the one that fits us best. 'It's allowing for this gene—environment correlation to operate,' says Hambrick.

In other words, we've got to be like Goldilocks. Try different chairs, porridge and beds, until you find the one that fits you best.

7

RUNNING

Exerting yourself to the fullest within your individual limits: that's the essence of running, and a metaphor for life.

Haruki Murakami

'I began running home from kindergarten when I was just six years old,' says Dean Karnazes. 'Running to me was freedom. It was a release, and a way to experience the world.'

At first, I thought this explanation of why a six-year-old from Inglewood in Los Angeles would be into running was suspiciously sophisticated. Why would such a young child need release? It's more likely, I thought, that Karnazes has transferred his adult understanding of running onto the motivation of his six-year-old self. He has grown up to become an extraordinary runner, after all. But later he mentions that he's a dyslexic introvert, and I start to think I underestimated his younger self. No doubt dyslexia can make childhood difficult and stressful, so I can see how running might provide freedom. I ask Karnazes if he felt different to his classmates. 'I didn't necessarily think I was different than other kids,' he says. 'My passions were just elsewhere, that was all.'

Well, even if he wasn't different then, he is now. And his

passion for running is second to none. Here's a selection of his achievements. On 12 October 2005, he set out on a run in northern California. He stopped only on 15 October, three days later, after running 350 miles. Karnazes spoke with a journalist from *Runner's World* magazine during the run.[1] From the transcription, a few moments stood out for me. At 3.29 a.m. on the Thursday, he reports seeing lots of skunk. And deer, bobcats, coyote and possums. By 2.21 a.m. on the Saturday, he realises he's been *sleep-running*. 'I suddenly woke up and realised I'm still running. And the really bizarre thing is that I feel like I got a little catnap.' By 9.07 p.m. he's closing in on 350 miles. 'Finishing this is as close to an out-of-body experience as I've ever had. Earlier on, the pain always brought my mind back, but for these last ten miles I've felt totally disassociated from my body.'

On 17 September 2006, he ran a marathon in St Louis, Missouri. Fine – but then he ran a marathon every day for the next forty-nine days, each one in a different state, ending in New York. Then – and this is getting preposterous – saying he wanted to clear his head, he set off to run to San Francisco. In the end he stopped in Missouri, after 1300 additional miles and twenty-eight days.

He's also won the Badwater Ultramarathon, which proclaims itself the world's toughest foot race. This is a maddeningly, monstrously difficult run through Death Valley, California – itself the record holder for the hottest temperature on Earth. The 135-mile race starts at 85 metres (279 ft) below sea level, and ends at 2548 metres (8360 ft) above, on the trail to Mount Whitney. Temperatures the day Karnazes won reached 49°C. For good measure, he's also run a marathon in Antarctica, in regular running shoes, at temperatures that dropped to minus 25°C. His name is routinely prefixed with the words 'super-human athlete'.

'I truthfully believe that anyone can do what I do if they have the same passion, drive, commitment and resolve,' he says. 'I think I'm simply excelling at something I love, which is no different than what many other people do in different fields.'

Love lifts us up (where we belong). Love spins you right round (baby right round). Love can carry you a long way, I get that. But with ultrarunners it's more – it's having a body that can propel you ever on while coping with incredible punishment, and a mind that can commit to relentless training and pain. If it's love it seems to be the kind that veers towards obsession, and a kind of addiction. I've never run further than to the bus stop, but don't get me wrong, I'm full of admiration for distance runners. It's because I can't do it that I want to understand how they can do what they do.

Like most people, I knew the etymology of the word 'marathon' – I knew that a marathon is 42.2 kilometres/26.2 miles because that is roughly the distance from Marathon to Athens, and that someone ran that distance to report on Greek success in a battle. Now I know, because a few ultrarunners have told me, that the someone is supposedly Pheidippides, who died in Athens after delivering his message.

The story goes on. A couple of days before the Marathon run, Pheidippides was sent with a message from Athens to *Sparta*, a distance of some 240 kilometres (150 miles), which took him two days. He then ran from the battlefield at Marathon back to Athens with his message of Greek victory over the Persians. It's a bit more understandable that you might collapse and die after that sort of accumulated distance. This, by the way, was in 490 BC – more than 2500 years ago.

But then I found out that there is no real evidence for any of this. The story of Pheidippides has probably been stitched together from snippets here and there and augmented over the centuries. No matter. It is such a good story that it ought to be true, and like all great stories, it has inspired millions of people ever since. If it wasn't true then – if the feats it tells are not historically accurate – it has been made true now.

Hundreds of marathons are held all over the world each year. Running is a passion for millions of people. For some, running a marathon is a one-off, something they want to achieve once in their lives. For others it becomes a habit, something more than a hobby, and for some a mere marathon isn't enough. For those who want to go full Pheidippides, there are dozens of ultra-marathons around the world, including the Spartathlon – a race that re-creates his supposed run from Athens to Sparta. This is a distance of 246 kilometres (153 miles). Another very impressive ultrarunner, 43-year-old Scott Jurek, has won the Badwater ultra and – seven times in a row – the Western States 100-Mile Endurance Run, as well as the Spartathlon. But the record for the latter race, 20 hours 25 minutes, is held by a legend in the world of ultrarunning named Yiannis Kouros. He has run the four fastest times the Spartathlon has ever seen, and he is sometimes called Pheidippides' Successor (another nickname is the Running God).

Extreme endurance running comes in two different classes, categorised by distance or time. For example, there are 100-mile road races, 1000-km track races and 1000-mile road races. On the other hand, there are 12-hour road races and 48-hour track races. Many of the men's records are held by Kouros, and of his achievements perhaps the most extraordinary is his 24-hour track record.

These races are mind-boggling just to contemplate. You run continuously, round and round a track, for twenty-four hours. The aim is to run as far as you can in that time. In

1997 in Adelaide, Kouros ran 303 kilometres (188 miles) in that period. That's an average pace of 7 minutes 39 seconds per mile. It's more than seven marathons strung together, without stopping, run at a pace not much over three hours per marathon. No one has come close to this record, and Kouros declared it would last for centuries. Someone should sequence that man's genome.

Nicole Pinto is an exercise physiologist at the Human Performance Center at the University of California, San Francisco. She's run a battery of tests on Dean Karnazes, including blood lactate and oxygen consumption tests. The thinking was that as he seems to be so far outside the realms of normal achievement, even for ultrarunners, there may be something interesting going on in him physiologically. She explains what happens in your muscles when you exercise.

Glucose is broken down into a compound called adenosine triphosphate (ATP), the unit of fuel used to power the contraction of a muscle. The byproduct of this is lactic acid, which breaks up to release hydrogen ions (hydrogen atoms that are positively charged because they have been stripped of an electron) and lactate. If there is oxygen present, the lactate can be converted back into glucose, and so used again as fuel. It's not the build-up of lactic acid itself that causes problems in muscles, as I'd thought and which seems to be the common wisdom – it's the accumulation of hydrogen ions. You've probably seen marathon runners going all wobbly as they completely run out of energy: the jelly legs are caused by hydrogen ions that acidify the tissue and interfere with muscle contraction. For athletes who rely on immediate strength and speed, such as weightlifters and sprinters, it doesn't matter if they produce too much lactate, because they are not going to operate at that full-on capacity for very long. For endurance athletes it's different, and for them the key is finding the balance between clearance and production of lactate. Better athletes can work harder for longer, because they are good at recycling the lactate that's building up in their muscles. They prevent hydrogen from acidifying their legs, basically.

'When we were looking at Dean we didn't see anything we'd never seen in other ultra-endurance athletes, but we did find he was extremely efficient with his sweet spot,' says Pinto. In other words, Karnazes has a much better ability to convert lactate back into glucose than regular distance runners. 'Dean's ability to stay in lactate balance is what's higher than normal,' Pinto says. His body doesn't find activity as strenuous as normal people.

Dean also has a highly efficient, economical style and maintains a consistent, steady pace. This steadiness may be the key to his success. 'His self-proclaimed style is "I don't run fast, I run really far",' says Pinto. 'So he is really effective at these long distances, and as long as he's replenishing his energy stores after hours and hours of running, he could run for ever.'

Once more, I want to assess how much of a particular trait is innate, and how much is picked up along the way. Karnazes says he thinks anyone can do what he does if only they have the passion and drive – but it's not like these things can just be decided on. Genetics affects your drive and dedication just as it does anything else. Remember how we saw, in Chapter 6, that much of the effort people put into practising can be explained genetically? No doubt similar genetic factors influence the effort you put into physical training. Despite media 'explanations' of Karnazes's ability as genetic,[2] he confirmed to me that no one has sequenced his DNA to look at his genetic make-up.

Jonathan Folland is reader in human performance and neuromuscular physiology at Loughborough University. 'In terms of how important genetic traits are compared to the environment,' he says, 'based on twin studies we know that inherited traits account for at least 50 per cent of the variability of physical ability between people.'

Assuming that figure applies to running, approximately half or more of how good you are at running would depend on your genes. Twenty years ago, this led to quite a lot of excitement among early researchers that we might find a 'running gene' that would explain a good proportion of individuals' ability. But just as with other complex traits such as intelligence, no such gene has been forthcoming. In fact there's not been a gene variant discovered that even explains 5 or 10 per cent of the variability between people.

'There's been quite a lot of work to uncover the specific genetic variables that might be important for physical performance,' says Folland, 'and that work hasn't drawn much of interest so far.'

It turns out that there are a few genes that maybe occur more frequently in Olympic-standard endurance athletes, but the importance of those genes is limited: 0.5 per cent, 1 per cent, maybe 2 per cent at most. 'It could be that there are 100 or 500 genes that each explain a fraction of a per cent,' says Folland.

Dean Karnazes brings up the Greek Marathon legend when I ask him about genetics and ultrarunning: 'I'm 100 per cent Greek and my father insists we're from the same village in the hills of Greece as Pheidippides.' The problem with this, as I'm sure Karnazes knows, is that even if he is related to a semi-mythical ancient runner, Pheidippides' genes will have been diluted over time almost to vanishing point in Karnazes.

More seriously, and despite his insistence that anyone can

do what he does, Karnazes accepts that genetics may play a role: 'There is a saying that the best thing you can do as a long-distance runner is to choose your parents right.' That certainly chimes with what Folland says.

Not only does his father possibly hail from the ancestral village of Pheidippides, Karnazes' mother's family is from the island of Ikaria. This has been named as one of Earth's Blue Zones: a place with a high proportion of centenarians, where people live longer on average than elsewhere, for genetic and other reasons. (We'll examine Blue Zones in depth in Chapter 8.)

Folland, who has coached athletes at international level, says there are three main physiological factors that determine endurance: maximum oxygen uptake (known as VO_2 max); something called fractional utilisation, which is essentially the proportion of VO_2 max that a person can sustain over a long period; and running economy, the biochemical and biomechanical factors that determine the amount of oxygen you consume when running.

'Those three factors combine together and explain a high proportion of the variability in endurance performance,' says Folland, 'whether you're talking about 5000 metres, 10,000 metres or 10 miles. Those three factors probably explain more than 80, 90 per cent.'

VO_2 max is typically measured as millilitres of oxygen per kilogram of body mass, per minute. The average healthy man has a VO_2 max around 35–40 ml/kg/min, the average woman 27–31 ml/kg/min. The difference is down to the larger physical lung size of men on average, and the fact that men have higher levels of haemoglobin in the blood. Basically, the higher your VO_2 max, the more oxygen you can deliver to your mitochondria, the energy-producing units in your cells, and the faster you can run. Training certainly helps improve your VO$_2$ max. Elite runners reach 85 ml/kg/min (men) and 77 ml/kg/min (women).

So if genetics explains a good half of running ability, and there are dozens, scores or even hundreds of contributing gene variants, it's quite possible that there are people who end up with a high proportion of them. There may even be whole groups of people who share these traits – people who would be if not born runners, then certainly well disposed to running.

The Sierra Madre Occidental is a range of deeply canyoned mountains reaching from Arizona to the west coast of Mexico. The intersection of geology and elevation brings rainfall to what would otherwise be arid land, and the region is noted for its high biodiversity, even nowadays in the face of anthropocene pressures such as mining and agriculture. Jaguars and ocelots, two dusky, beautiful wild cat species, are still sometimes seen. The Mexican wolf is here too, but limited to the southern part of the range. These mountains are also the home of the indigenous Tarahumara people, who call themselves the Rarámuri – the name means something like 'foot-runner' or 'those who walk well'.

Sounds romantic, right? It does, when put like that. Add that the reclusive Tarahumara are a people with a culture of running, and that some of them can do it really well, and you've got the makings of a potent myth. In 1993 a previously unknown Tarahumara runner named Victoriano Churro won one of the more gruelling ultramarathons in North America, the Leadville Trail 100. He was fifty-two at the time, and it was the first time the Tarahumara had raced outside Mexico. Just to clarify why the Leadville is more gruelling even than a regular ultramarathon, its route, in Colorado, goes through the Rocky Mountains and runners slog up and down 4800 metres of elevation over the 100-mile course. The drop-out rate is high: around half of the starters fail to finish within the thirty-hour time limit. The following year, 1994, another Tarahumara runner won the race. After that they disappeared,[3] but their legend was established.

The journalist Chris McDougall brought the Tarahumara to the world's attention in his book *Born to Run*, and the stories about them began to snowball, particularly in the running community. Since they tend to run in very basic sandals called *huaraches*, which are simply flat soles tied to the runner's foot with straps, the Tarahumara have inspired a whole movement of so-called natural running – either barefoot or wearing huaraches, or minimal footwear such as Vibrams Fivefingers, rather than cushioned running shoes. The Tarahumara use pieces of leather or car tyres for the sole. McDougall argues that cushioned running shoes contribute to foot-strike injuries, and that the footwear (or lack of it) of the Tarahumara is much more conducive to healthy running.

He has the support of Daniel Lieberman, an evolutionary biologist at Harvard University, who studies the biomechanics of endurance running, with particular regard to running barefoot. Such is the interest from the running community, he created a website which presents the pros and cons of barefoot and cushioned running.[4] Wearing shoes completely changes the way you run, says Lieberman. 'By landing on the middle or front of the foot, barefoot runners have almost no impact collision, much less than most shod runners generate when they heel-strike. Most people today think barefoot running is dangerous and hurts, but actually you can run barefoot on the world's hardest surfaces without the slightest discomfort and pain. It might be less injurious than the way some people run in shoes.'

A study specifically of the Tarahumara found that the type of shoe changed the way their feet struck the ground. When running wearing huaraches, the forefoot and the midfoot struck the ground first 70 per cent of the time, and the hindfoot struck first the remaining 30 per cent, but when wearing cushioned shoes, the hindfoot struck first 75 per cent of the time.[5] If you are not a runner this might all be puzzling, but hindfoot strike is a big deal, because it is linked to a variety of repetitive stress and other injuries, such as Achilles tendonitis. Indeed it was a foot injury that started McDougall's investigation into the Tarahumara.

This is all very measured and sensible. But there is an unfortunate tendency to romanticise indigenous cultures. According to one account, the Tarahumara have 'a diet and fitness regimen that has allowed them to outrun death and disease'.[6] A documentary about them claims they don't have cancer, diabetes or hypertension.[7] Other reports state that the Tarahumara are able to run huge distances in one go, delivering messages between widely separated villages, or to run down deer.

Make no mistake, some Tarahumara can certainly run well. At the first World Indigenous Games, which took place in Brazil in 2015, Tarahumara runners came second and third in the 10,000 metres.[8] In the 2016 Copper Canyon Ultramarathon, a 50-mile race founded by a legendary runner named Micah True, Tarahumara runners took the top three places. Yes, they have ability – the question is, are the Tarahumara *innately* talented runners? It's a key question: for those who can run well, how much is genetic, and how much is trained?

Dirk Lund Christensen, a physiologist in the Section of Global Health at the University of Copenhagen, has studied the health and fitness of the Tarahumara first hand, after coming across them by chance in Mexico some twenty-five years ago and becoming fascinated by their culture of running.

In 2011, Christensen and colleagues organised a 78-kilometre (48-mile) race to test scientifically the running prowess of the Tarahumara. His team recruited ten men from the village of Choguita, Chihuahua, and briefed them on the task. They were to run a flat loop of 26 kilometres (16 miles) three times, starting from a hospital, the Centro Avanzado de Atencion Primaria a la Salud, in the town of Guachochi at an altitude of 2400 metres (7900ft). At 5.55 one November morning, the runners, all but one wearing homemade huaraches and loincloths, set off. If you believe the hype that the Tarahumara have innate talent, then runners would sail round – they would float round. After all, 78 kilometres should be a breeze for them.

'About half of them had difficulties completing the race and they had to walk a considerable part of the distance,' says Christensen. The scientists took blood from the runners before the race, immediately afterwards, and at several points in the following hours and days. They measured the runners' VO_2 max and other parameters such as blood pressure. Since we've learned about VO_2 max, it might surprise you to find that the average of the Tarahumara runners was 48 ml/kg/min – not very high, says Christensen, for runners capable of completing an ultra-distance race of 78 kilometres.[9] Elite Western runners, remember, can have a VO_2 max of 85 ml/kg/min. If the Tarahumara were innately skilled runners, you might expect higher. Clearly, the Tarahumara who win races are trained athletes.

In another study, Christensen and colleagues tested sixty-four adult Tarahumara for cardio-respiratory fitness. They found that hypertension and diabetes do indeed exist in the Tarahumara.[10] Sadly, for the myth of a super-healthy tribe, but more for the people themselves, the obesity epidemic that has swept the world has not left them untouched. 'Studies going back more than fifteen years have shown that obesity is a considerable problem in Tarahumara women,' says Christensen. 'An epidemiological transition has been going on for quite some time.' A transition, that is, from a lifestyle based around running to a sedentary one.

'The results,' says Christensen, 'clearly show that being a Tarahumara does not guarantee superhuman running abilities.'

Christensen bemoans the impression given by those in the running community, and some academics, that the Tarahumara are free from cardio-metabolic diseases. 'A look into hospital records from the area where the Tarahumara live as well as scientific studies show that this is untrue,' he says.

And it's not just obesity that threatens the Tarahumara's well-being. In 2015 the Copper Canyon Ultramarathon, held in Tarahumara lands in Mexico, was cancelled because of the danger of drug violence.[11] Much of their land borders the Golden Triangle, an area of intensive heroin, poppy and cannabis cultivation, and drug cartels have been actively targeting young Tarahumara men to act as traffickers.[12]

'Maybe we in the Western world have an urge to glorify people who in some ways live like we did many generations ago,' says Christensen, 'a "pure" life free from chronic disease and with exceptional physical stamina due to lack of motorised transport.'

Lieberman's work may inadvertently tap into this yearning. As an evolutionary biologist, his perspective is that we didn't wear cushioned running shoes for the vast majority of our history, so those rearfoot strike injuries were probably far less common. It may be true that the huarache-wearing Tarahumara suffer fewer of these injuries, but as we've seen they're not free of other modern diseases.

There are superb Tarahumara athletes, that's plain to see, but is doubtful that the Tarahumara as a people have genetic traits which make them more talented as runners, certainly not without training. Christensen hasn't tested this, but he plans to. 'It is much more likely,' he says, 'that those who are talented – assuming a relatively normal distribution of running talent in this particular population – are also very active and thereby optimise the ultra-distance running talent.'

When I started reading about the Tarahumara I was just as swept up by the romantic myth as many others. I looked a little more closely because I'm trying to understand how the best can do what they do. Rather than having a mythical floating style of running, it seems that running is just something that the Tarahumara do. The psychological impact of growing up in that culture – social facilitation, the professionals call it – goes a long way towards explaining why ultrarunning is the norm. Or was. The sad thing is that their culture is falling away, and the ills of modernity – a sedentary life, with associated diabetes and hypertension and obesity – are replacing it.

So to summarise what we know about the genetics of running, both for individuals and for athletes in general, we have clear evidence that genetics plays a role in determining how well people perform during training exercises. That's non-controversial. What is more shaky, however, are attempts to characterise *groups* of people in the same way.

Since East Africans started dominating long-distance events in the 1990s, there have been several attempts to attribute their success to genetics; to say, in other words, that East Africans as a group are genetically better equipped. The problem is, no group-wide genes have been found that could explain improved performance.

It's more likely, then, that living as the East African runners do, at altitude – the Kenyan capital Nairobi and the Ethiopian capital Addis Ababa are respectively 1800 and 2300 metres (5900 and 7500ft) above sea level – increases red blood cell count, giving them a big boost when running at lower levels. The tendency to be light and lean, as many East Africans are, is also a big help for efficient endurance running. 'The physiological evidence so far,' says Folland, 'is that East Africans have a better running economy, so they are more efficient in terms of the way they run. Part of that is anatomical – they're slim, which is crucial, as it reduces limb inertia.' This is basically the energy expended in waving your arms backwards and forwards as you run. Folland says that some of the East African athletes who perform at the very highest level may have other 'structural' advantages, such as a longer Achilles tendon. Runners with long Achilles tendons have calf muscles that attach relatively closer to the knee, which reduces the inertia of the leg and so improves running economy. 'A longer Achilles tendon is potentially also better for storing energy,' Folland says. Like the Tarahumara, there's also a big element of social facilitation: 'A lot of the rural kids run to school and back every day so you have this big pool of children who are very active and trained before they even get into competitive sport.' In both the Tarahumara and East African runners, there is also a big source of social motivation. Winning races can provide prize money that goes a long way towards helping a village in a poor country. It's like we saw in Chapter 4, with the effect of motivation on focus.

'Stamina is certainly a trait in the Tarahumara, but it comes about from mental will power as much as physical stamina,' Christensen says. And so we're back to the role of psychology. Just how far can it get you?

*

I'm waiting to talk with Petra Kasperova in a running shop in London. At first glance it's a regular running shop, with shoes and clothing and energy drinks in neat racks, but there are also framed photos on the wall of an Indian spiritual leader, Chinmoy Kumar Ghose, posing with various legends of athletics and sport. There he is with Carl Lewis. Muhammad Ali. Paula Radcliffe. There's also a picture of him with a lesser-known figure, Tony Smith, who founded the shop, Run and Become, in the 1980s. Sri Chinmoy (who died in 2007 and who is usually referred to with the honorific Sri) was born in East Bengal, India (now Bangladesh), and moved to New York in the 1960s, where he started a meditation centre which combined meditative practice with athletic development. There was no point, he taught, in just concentrating on your inner life if you ignore the outer case that it's in. The outer case: I like that. It makes me think of a caddis fly larva inside its stony home, a case designed by the animal's genes, and put together using what it finds in the environment; it makes me think of Richard Dawkins' description of us as lumbering robots controlled by our genes.

Running, Sri Chinmoy taught, is an opportunity to challenge what you think you're capable of. It's a means to expand the limits of human potential. He started a marathon team and a running club, and his ethos spread. There are now hundreds of Sri Chinmoy events around the country and the world. So that's why I've come here, to a shop started by one of Sri Chinmoy's disciples, to talk to people who take his message seriously.

Like all the staff in the shop, Kasperova, twenty-seven, is dressed in running gear. She doesn't particularly stand out; she's ostensibly a normal young woman. But a few weeks before we meet, she completed one of the toughest challenges in world sport, a six-day ultramarathon. The Sri Chinmoy six- and ten-day races take place each year in Flushing Meadows, Queens, New York City. This is the home of the US Open tennis tournament and of the New York Mets baseball team: Flushing Meadows is already a hotspot for extremes of sporting glory and emotion. But in the park too is a one-mile looped running track. The track is flat and the route scenic. The race is simple, and devastating. The ten-day event starts at noon on a Monday; the six-dayer starts four days later. Both races finish the following Thursday, at noon. The winner is the person who has run the furthest. In April 2017, Kasperova took part in her first multi-day race, starting with the six-dayer. She ran 64 miles on the first day, then 62, 52, 48, 45 and finally 48 miles. The last three days, she says, she was hampered by injuries and she had to walk some of the way. But the total of 319 miles placed her fourth of the women runners. To put it another way, she ran more than twelve marathons in six days.

I realise I've been thinking of ultrarunning as a macho thing. I've marvelled at the records of Yannis Kouros. I've interviewed super-toned legend Dean Karnazes. I've been influenced by photos I've seen of sweaty, muscly, grimacing men pounding topless up mountains or through deserts or across ice fields. My idea of ultrarunning has been swamped, in short, by the testosterone-laden Ironman brand. But marathon runners aren't bulky, they are slender.

When she was a child in Prague, Czech Republic, Kasperova says, she was always active. She loved to be outside, and still does. She loves being connected with nature, being under the sky, and feeling the sensation of fresh air. But as she got a little older, her friends started slowing down. 'During my teenage years, I was surrounded by people who seemed to be miserable, so I covered myself a little bit,' she says. She withdrew a little, she hid her nature from herself, she was unsure of herself. Then when she was nineteen her parents went through a divorce, and she had her final university exams. 'I was completely lost, I couldn't sleep. I had such headaches, I felt I couldn't do my finals. There were so many questions in my mind. Suddenly I thought: I need to silence my mind.'

Searching online for free meditation classes in Prague, she found a Sri Chinmoy practice. 'I found this class and it changed my life, something just clicked. This resonates with my heart. I felt, I don't have to be like all these people in my class.'

Following the Sri Chinmoy way, she took up running. This was six or seven years ago, around 2010. She started gradually, with short distance running, 5Ks and 10Ks, then her first half marathon. She soon built up to a marathon, then an ultra. To my mind that's already extraordinary, but then she ran the 24-hour Self-Transcendence race in Tooting, south London. The race is contested by forty-five invited runners, who circle the Tooting Bec athletics track as many times as they can, and is organised by Tony Smith's daughter, Shankara. 'Some people say the purest kind of ultra are the races on the track,' she says, referring to the long-distance races on a looping track where you just go round and round for ever, or for what feels like for ever. In these events, there's nowhere to hide. 'It's only you, you're facing your demons and what you think you're capable of. If it's you against some tough terrain you can fight the mountain, but on the track it's just you.'

After this, Kasperova entered the six-day race at Flushing Meadows. Tents spring up in the running village that appears during the race. Physio tents, food tents and sleeping tents. There are large communal sleeping tents, but Kasperova puts up her own. She wanted somewhere she could have her own space, somewhere she could crawl into and cry at the end of eighteen hours of solid running. Whatever time she finished, midnight or one or two in the morning, the alarm was set for 5.45 a.m. and she was up and off again.

'You always want to know what you are capable of,' she says. 'What human nature is capable of. We have so much inside ourselves. And you learn so much about yourself. When you see how much these things change your life, you want more. You want to know more.'

I must say, talking to Kasperova I feel almost transported myself. At her young age she has the calmness and serenity of a guru. I'm not a runner – but this makes me want to start. She says that she feels lighter after a race. She feels that after a race she's become a different person, that her nature has changed, in a positive way, and that she has experienced a beautiful form of happiness. I know this is the sort of thing that has non-runners rolling their eyes – I've done it myself – but in the face of it any cynicism I have falls away, and is revealed in fact as a feeling of inferiority compared to people who do these things. (Only a mild feeling, mind. Perhaps admiration is better – you can admire someone without feeling inferior.) She also runs for those who can't. A friend died recently of cancer. 'She would have loved to do it. Even though I was in pain during the six days I just wanted to do it for the people who can't. It's about keeping going and never giving up.'

Doing these extreme events, something happens, Kasperova says. 'Something magical. You want to do more because you want to experience these feelings of absolute happiness and absolute peace that you don't get in the normal world. It just fills up your heart and it's hard to get that in the day-to-day routine.'

I watch some race footage of the runners taking part in the six- and ten-day races. Sure, in the film they don't include the tears and the agony the runners go through, but what jumps out is the smiling, joyous nature of the people running around. Kasperova says she feels the energy circling her body. 'You've had two hours' sleep and you're running 60 miles a day. I don't think it's really me, it's something higher.'

Who would have thought that this track in Flushing Meadows, squeezed between Interstate 678 and LaGuardia Airport, humming with the sound of traffic and keening aircraft, could be the setting for such, well, transcendent joy? And who would have thought that in the runners circling the loop, psychological battles would be raging every bit as intense and impressive as those flashier ones played out on the tennis courts or the baseball pitch nearby?

Kasperova is not done with the six-day race, as she feels she has more to learn from it, and after that she wants to do the ten-day race. 'And then someday the ultimate.'

The ultimate is Sri Chinmoy's final creation. It's hard to describe it without coming to the conclusion that he was an evil genius. After all, by the time he was creating the really long races, his knees were shot and he could no longer run them himself – but he kept creating longer races for his followers to run. He took up weightlifting, which he excelled at, but devised ever more torturous races. Sri Chinmoy, says Shankara Smith, was all about challenging things considered impossible – about unlocking potential that we don't think we can reach. And no one's forcing people to attempt these ridiculous races. The ultimate is the Self-Transcendence 3100-mile race. It also takes place around a route in Queen's, but not the Flushing Meadows track – this is around a city block, 5649 times. With running, says Shankara Smith, it's about you against yourself. 'It's your self-doubts you have to tackle, and the longer the race, the more it's about challenging your perceived impossibility level than the whole physical fitness aspect.'

Runners have fifty-two days to complete the race, which amounts to over 59 miles a day. The current record-holder is a Finn, Ashprihanal Aalto, who in July 2015 completed the 3100 miles in 40 days, 9 hours, 6 minutes and 21 seconds. Astonishing, but it's not even the most punishing Buddhist-inspired event I've heard about. That surely is the *kaihogyo* route around Mount Hiei near Kyoto, Japan, the most extreme version of which involves running 1000 marathons in 1000 days. Few people have ever achieved this feat.

I mentioned Sri Chinmoy's advice, about taking care of the 'outer case' – our bodies. But he knew that it's more than just taking care of the outer case. If you push the shell to extremes, something feeds back into the mind and changes it, and that's why it's a transcendent experience. But no one imagines that there is literally a line you cross when transcendence happens. It's a continuous thing, which is why you always want to go further, push harder. Genshin Fujinami, one of the few monks who has completed the '*kaihogyo* 1000' has said that for him, the quest continues. 'The 1000-day challenge is not an end point, the challenge is to continue, enjoying life and learning new things.'[13]

I said earlier that Kasperova was ostensibly a normal young woman. She would say she *is* a normal young woman. That's her point – that anyone could do what she does. 'I think most people could do this 24-hour race,' she insists. But there's a huge caveat. 'If they have the will power.'

Dean Karnazes told me how he used to run home as a kid, and I could sense the passion of his six-year-old self. And I started this chapter by listing some of his more recent feats. But what's odd is that for a man who now so embodies running, for whom it is such a central part of his life, during all of his twenties he didn't run at all. He was a young man with a job and a social life that was built around alcohol. Nothing unusual there, but on his thirtieth birthday, doing tequila shots with his friends, he had an epiphany.

'I felt as though I'd followed the societal prescription for happiness – get a good education, get a good job, have a comfortable and secure future – yet something was missing,' he says. 'Instead of feeling happy I felt empty, as though I'd betrayed myself. It all came to a head on the night of my thirtieth birthday when I walked out of a bar, three sheets to the wind, and decided to run 30 miles in celebration of my thirty years of existence on planet Earth.'

Karnazes hadn't run for over ten years.

'I ran straight through the night without rest. The aftermath wasn't pretty, but it didn't matter. I'd accomplished what I'd set out to do; the blisters and chafing and muscle cramping didn't matter at all. My destiny was now clear. That night forever changed the course of my life.'

It does sound like a superhero origin story, doesn't it. And indeed Karnazes appeared on Stan Lee's *Superhumans* TV show, which is why he ended up being tested at Nicole Pinto's lab. I assume he is happy now he's running again, or at least happier than he was in a bar downing tequila. A friend who has run thirteen marathons says running has made her more optimistic, and makes her feel more alive and more healthy; it also helps her get through other problems in life.[14] 'When I am actively running, in particular through somewhere beautiful, I feel intense joy, and like all other emotions are enhanced,' she says. It's got to be worth a try.

Part III

BEING

8

LONGEVITY

For age is opportunity no less
Than youth itself, though in another dress,
And as the evening twilight fades away
The sky is filled with stars, invisible by day.

Henry Wadsworth Longfellow

At 70 you are still a child. At 80, a young man or woman.
If at 90 someone from Heaven invites you over, tell him:
'Just go away, and come back when I am 100.'

Saying carved into a rock facing the sea near the village of
Kijioka, in Okinawa, Japan

As I write this, I am 16,931 days old. My life expectancy comes
in at 30,736 days. So, to a good approximation, I have 13,805
days left.

In more conventional notation, that means I'm forty-six,
can expect to live until I'm eighty-four, and have thirty-eight
years left. But reckoning your life in days is a quickening thing
to do. Try it. Calculate how many days you've already lived.
If you're like me you get a feeling like sand slipping through
your fingers. You'll probably think, oh my God I should do

more with my days. Don't worry, this feeling will pass, and you'll carry on living your life as usual. There aren't many people who can live a breathless, *carpe diem* life. It's just too much effort, constantly seizing the day. What we'd like, what we've wanted for thousands of years, is the promise of limitless days, so we can waste them as we like, but use them as we like too. I don't want much, I won't even ask for immortality, just more than my predicted 30,736 days. And I'd like them to be healthy days.[1]

Elizabeth Love has already lived for 37,164 days. She's 101 and is the oldest person I've ever met. But centenarians are becoming more common. In 2000, there were an estimated 180,000 centenarians worldwide and the United Nations says there will be 3.2 million of them by 2050.[2] Life expectancy has been increasing for decades now. It's why my daughter has a longer life expectancy than me, coming in at just over ninety-five years. As I write, she's only lived 1546 days, and has a projected 33,180 left. In Japan life expectancy is even higher: a child born in Japan today has more than a 50 per cent chance of living to be 107.

That's all very well and good for them, isn't it. But we want that longevity here and now. We want to have youth and vigour now and into our old age. Many scientists and research institutes are working towards this. One, the Methuselah Foundation, based in Springfield, Virginia, aims explicitly to slow the clock. By 2030, it says, technology will have us looking and feeling fifty when our chronological age is ninety. Ageing, to a growing number of scientists, is a disease – a genetic one we all suffer from. It is the number one cause of death, because all the diseases we die from – cancer, cardiovascular disease, dementia, diabetes – are those sprung from a common cause: old age. Old age is a disease we can cure.

You might not care about bravery, you might not care about singing ability. You might not even care much about intelligence. But you surely care about how long you will live. In this chapter we'll look at how lifespan has been increasing, and what we're learning about the factors behind it. We'll meet some very old people, and see what we can learn from them. I'm far from the only one doing this – centenarians around the world are being studied in the hope that we can glean their secrets, and have some long life for ourselves. But as I meet them I'm going to bear in mind a fact that no one seems to mention in the hundreds of books and articles claiming to reveal the secrets of living to a hundred, or how to eat your way to that age. It's this: few of the centenarians, if any, *tried* to live to a hundred. They just did it. They weren't pill-popping or calorie-restricting or whatever – they just blundered through. Let's see if we can find out how.

Elizabeth Love was born in 1915 so she has lived through two world wars, but, as Gandalf once said of Bilbo, she seems extraordinarily well preserved for someone 101 years and 269 days old.

Her age, she says right away, must be something to do with having good genes. Longevity runs through the family. Her own grandmother was ninety-three when she died. 'My mother was the youngest of three sisters,' she tells me. 'She was eighty-four when she died, the second sister was ninety-three and the eldest sister was 101. My cousin, the son of the middle sister, he only recently died, at ninety-six. All the same genes, there must be something in that.'

There must indeed. We feel, just from knowing our family histories, that longevity is to a certain extent inherited. Turning that feeling into an understanding, however, has been tricky.

Much of our understanding of the heritability of lifespan comes from an influential and long-running Danish twin study. As we've seen in earlier chapters, twin studies are invaluable to biologists trying to tease out what is genetic and what not. Since identical twins share all their genes and fraternal twins half their genes – and because all have almost the same environment – it's possible to examine a trait such as lifespan and determine how much is influenced by genetics and how much by other things. The first big paper on this, which came out in 1996, looked at 2872 pairs of twins born between 1870 and 1900 and followed them until the mid-1990s, when almost all of them had died. The analysis showed that genetics accounted for 26 per cent of the variation in lifespan for men, and 23 per cent for women. In other words, not a huge amount.[3]

For a while that seemed to shut the door on any search for longevity genes. Genetics didn't seem to have much to do with it, and it seemed that lifestyle and environment were far more important. That feeling still persists among some gerontology researchers I've spoken to, and indeed no one will dispute that these non-genetic things are important. But ten years after that study, Kaare Christensen's team, at the Institute of Public Health, University of Southern Denmark, took a closer look at what was going on with the very old, and noticed something interesting.

The Danish twin data set was now a bit bigger, and extended to people born up to 1910. This gave the researchers 20,502 individuals to look at. Christensen's team now found evidence that genetics is much more important than previously thought in determining whether you live into your nineties and beyond. 'These findings,' the authors wrote in the journal *Human Genetics*, 'provide support for the search for genes affecting longevity in humans, especially at advanced ages.'[4]

So the search is on again. To call it a gold rush might be going a bit far – but not much. For thousands of years people have yearned for a longer life. In Hindu mythology it was *amrita*, the nectar of immortality; in ancient China the elixir was thought to contain (disastrously, for the emperors who drank it) mercury; in medieval Europe the Philosopher's Stone was a substance supposedly able to grant eternal life. Now the objects of fascination and study have names such as *ApoE*, rapamycin, *FOXO3a* and others, as we shall see. The search for the factors linked to lifespan is the modern equivalent of the historical search for the fabled elixir of life.

When I arrive at Mrs Love's spacious and warm flat in Beaconsfield, about 25 miles north-west of London, it's 2 p.m. Her daughter has told me that her mother fell and broke her hip earlier in the year, so I'm prepared for someone quite frail, but Elizabeth Love is standing up to greet me, and looks, remarkably, at least twenty years younger than her chronological age.

She asked that I arrive after lunch. This is after, I now find out, her daily glass of sherry. 'I have a glass of sherry before lunch every day. And I have a gin and Martini every night before my supper. And have done for years.' There's just a hint of defiance. (I should've arrived *before* lunch, I think.) She had other vices, too.

'I smoked all my life, not heavily, but I did. I gave it up about ten years ago.'

This wasn't on doctor's advice, she just gave up, just like that.

'I smoked about ten a day, so I wasn't a terribly heavy smoker but I did smoke and I still drink.'

Ten cigarettes a day for the best part of seventy years seems like quite a lot to me, and any doctor will tell you that for most people, the vast majority, smoking like this will see you off well before you reach old age. And yet. You hear of people who can get away with it. Mrs Love appears to be one of those herself. Jeanne Calment, the most famous person in ageing research and the oldest human to have ever lived, was another. She apparently smoked for ninety-six years.[5] She is not, however, thought to have smoked more than one or two cigarettes a day, nor is it known if she even inhaled.[6] Still, even very light smoking is harmful, so she must've been protected, somehow. She got away with it.

Calment lived her whole life in Arles in the south of France, dying in 1997, aged 122. She ate a diet rich in olive oil and chocolate: she reportedly put away a kilogram of chocolate a week. We feast upon factoids like this, grasping for justification for our own bad habits. When I casually mentioned that Mrs Love had a glass of sherry before lunch every day, my partner immediately said no, I couldn't use Mrs Love as justification for drinking.

Inevitably, Mrs Love and I start off talking about the Second World War. She lived on Kensington Church Street in London until her flat was bombed in the Blitz. 'The blast blew the windows out,' she says. 'I wasn't able to go back then, much to my disgust.' Her husband was a South African who enlisted in the navy then worked for the Admiralty. He was three years younger than her, and died in 2004.

'We got married in 1941 when he was home on leave,' she says. Their plan had been to marry in September, but in July her fiancé was told he was going to be posted abroad. They had no option but to arrange a snap wedding for the next Saturday. The day after they married, he was posted away with the navy for eighteen months. 'Our communication, my husband and I, was through air letters. I was lucky if I got one every three weeks. And that was the only communication we had for eighteen months.'

Mrs Love joined the Women's Voluntary Service. She had learned to drive before the war, so she was tasked with delivering supplies to those homeless after bombing raids. 'I was driving an army lorry, we went to Coventry, when Coventry was virtually flattened. We were at the Blitz in Liverpool. I was driving a water carrier, and we delivered food to the people.'

By the end of the war she was living back in London with her husband. 'We were in the crowd outside Buckingham Palace, on VE Day,' she says. 'We saw the Royal Family come out.'

She doesn't really have friends, now. 'When we first came here I used to play bridge with people in the flats, but one by one they gave up.' She corrects herself. 'They died, let's face it. And so it all died a death, so I don't play bridge any more.'

Her daughter interjects. 'You've never done any exercise, have you, Mum?' Mrs Love happily agrees she's never bothered with anything like that. It's a counterintuitive point that we'll come back to: on average, centenarians don't do any more or less exercise than the rest of the population. Not all have engaged in a healthier lifestyle, nor have they all set out to be long-lived. Swapnil Rajpathak and colleagues at the Albert Einstein College of Medicine, in New York, surveyed 477 exceptionally long-lived people (their average age was ninety-seven) and compared them with a selection of younger people from the general population. All the participants were asked about their lifestyles. Rajpathak found that the super-old group were just as likely as younger members of the population to be overweight or obese, had similar patterns of alcohol consumption and physical activity (or lack of it), and were just as likely (or not) to have a low-calorie diet.[7] The environment, what they do or what they eat are not factors for centenarians reaching old age.

I ask Mrs Love if she's lonely, and she says no, although her daughter is sat with us and the denial is not wholly convincing. She's not worried about loneliness, she says. Every day she does the *Telegraph* crossword, goes for a walk with her carer and chats with someone. They all know her in the shops. One of her daughters visits almost every day. 'I've got eight grandchildren and ten-and-one-on-the-way great-grandchildren, so it's a big family. They've kept me young. It keeps me going, keeps my interest in life, I listen to all the tales of the family in various age groups and generations, and it definitely keeps me ticking over, it gives me a huge interest in life.'

She's had to broaden her views. 'I've had to accept things I don't approve of,' she says, and I brace myself for some awful and anachronistic view. 'This boy and girlfriend living together thing, I don't approve of that out of marriage.' That's as bad as it gets. 'There's no doubt about it,' she says. 'Having a big family has kept my brain ticking over.'

No one has yet beaten Jeanne Calment's longevity record. The closest is an American, Sarah Knauss, who died in 1999 aged 119. Knauss and Calment mark the current crest of a remarkable surge in life expectancy over the last 200 years, a surge that has encouraged wild optimism about how far it might go.

In the nineteenth century, you could expect to live to somewhere between thirty and forty. People did live longer than that, of course, but because it was common for children to die young, the average life expectancy was dragged down. But in rich countries, for every year over the last two centuries, three months has been added to life expectancy. At the same time, people are having fewer babies — fertility has crashed throughout most of the world. The proportion of older people to younger is shifting.

The shape and composition of human society is changing

more rapidly than at any point in history. The gigantic shift in demographics will have profound implications that governments are only just waking up to. It raises the question of whether the increase in longevity will just carry on.

Jan Vijg, of the department of genetics at the Albert Einstein College of Medicine in New York, thinks not. In 2016 he and his colleagues published a paper in *Nature* suggesting that we'd reached the natural limit of the human lifespan.[8] Sure, our life expectancy has been rocketing, they said. But now it's plateaued. It's one thing to show how the lifespan of nematode worms or lab mice can be extended, but quite another to extrapolate to humans. The steepness of the life expectancy curve has beguiled us. By analysing births and deaths in these databases of lifespan from around the world, Vijg's team found that the age at which the oldest person died each year increased gradually throughout the twentieth century, but juddered to a halt in the nineties. No one, they found, seems to make it much beyond 115. The plateau in the rise in lifespan is evidence that humans, like all other animals we know of, have an upper limit to their age. There is a biological ceiling, and we've reached it.

But the conclusion was immediately attacked. James Vaupel, director of the Max Planck Institute for Demographic Research, in Rostock, Germany, pointed out that lifespan records have often been broken, and he robustly criticised the quality of the Vijg analysis, saying, 'They just shovelled the data into their computer like you'd shovel food into a cow.' A rebuttal paper by Stuart Ritchie (who we met in Chapter 1) and other scientists, published in *Nature* in 2017, attacked the statistical methods used in Vijg's paper, and dismissed the conclusion.[9]

So whether or not Jeanne Calment reached the natural limit of our species' lifespan remains to be seen. In a way, many researchers don't care: they are an optimistic bunch, and are forging ahead to find ways to lengthen our healthy lifespan, regardless of any potential 'natural' limit. 'Attempts to intervene in ageing in general are going to be frustrated by complexity, but this is not a reason to not try,' says Jay Olshansky, a leading researcher on human ageing, based at the School of Public Health at the University of Illinois, Chicago.

Lifespan is an incredibly complicated trait, and to push it to remarkable levels will take some extraordinary interference. We know that some exceptional individuals, such as Elizabeth Love, reach extreme old age without suffering debilitating disease: they are protected, somehow. We know we won't be able to keep pushing the top end of life upwards just by eating well. We'll need to find out more about what centenarians have in common. Fortunately, if you look at demographic data from around the world, you find that centenarians tend to cluster in certain areas. These are the longevity hotspots otherwise known as Blue Zones.

Drive up Highway 58, the seaside road on the western side of the main island of Okinawa, in the far south-west of the Japanese archipelago, and head north towards the jungle zone. On your left are subtropical waters of corals and bright reef fish and even a few dugongs, curious marine mammals also known as sea cows. The north of the island, Yanbaru, is a heavily forested jungle, rich in biodiversity. At dusk you see giant fruit bats, and, if you're lucky, the Okinawa rail, a rare flightless bird (once when I was there I *think* I caught a glimpse of one). The *habu*, a much-talked-about venomous snake, lives in the jungle, and I definitely saw some of these when I was working there as an insect biologist, although you're more likely to see one coiled in a bottle of the local *awamori* sake than on the path in front of you. Pickling the snakes in alcohol is supposed to imbue the drink with stimulating properties.

What you will see, especially in the village of Kijioka, are super-sprightly octogenarians. That's remarkable enough, but these old people are considered youngsters by the centenarians here. I met some of them, but it was hard to communicate as they spoke the Okinawan dialect and I spoke only standard Japanese. Still, there's something amazing about a place where an eighty-something is called, in perfect seriousness, a youngster. One elder, a 105-year-old woman from Kijioka called Nabi Kinjo, became famous when she encountered a *habu* on her porch. She clubbed it to death with a fly swat.

Japan has consistently topped the world longevity charts for years. It helps that every town and village in Japan maintains a *koseki*, a register of births, deaths and marriages dating back to the 1870s. The data in these *koseki* support the claim that long life in Japan isn't a recent phenomenon. Sure, the average lifespan has been increasing in Japan as it has elsewhere, but there's something that puts the Japanese at an advantage.

Of course not everyone in Japan lives to a great age. There is variation, and there are hotspots where individuals enjoy an especially long life. But within the country, it is the people on the island of Okinawa who live the longest, and within Okinawa it's the people of the village of Kijioka. The people on Okinawa's western edge are the longest lived in the world.

The people here have been well studied for clues to their remarkable lifespan. Diet (high in tofu, fresh vegetables and fresh fish), social structure (tight-knit, supportive), lifestyle (activities such as *bashofu* – a traditional form of fabric weaving that keeps people occupied into old age and supposedly maintains cognitive health) are all considered factors. As is the habit of *hara hachi bu*, a Confucian practice of eating only until you are 80 per cent full.

Several genetic studies have been conducted too, as they have among various other groups of centenarians. Unsurprisingly, Okinawa has been declared a Blue Zone.

As well as Okinawa, I'm lucky enough to have been to some other Blue Zones – Sardinia, in the Italian Mediterranean, and the Nicoya peninsula of Costa Rica – and I can appreciate why their environment could be so conducive to a long life. They are warm and comfortable, food is healthy and abundant. Another, as we saw in Chapter 7, is the Greek island of Ikaria, home of ultrarunner Dean Karnazes' mother. Surprisingly, there's a Blue Zone in the continental United States, the homeland of fast food and the epicentre of the obesity epidemic: it's the city of Loma Linda, in California.

Loma Linda was classed as the only Blue Zone in the US by Dan Buettner, an explorer and writer who came up with the concept after researching longevity hotspots around the world. Men in Loma Linda have a life expectancy of eighty-eight, and women a year more. It's a lifespan some eight to ten years longer than that of the average American, and the reason is fairly straightforward: the town has been extensively settled by members of the Seventh Day Adventist Church. Seventh Dayers don't drink or smoke (smoking is banned in the town) and most don't eat meat. The religion strongly encourages exercise and healthy living, and members of the church also regularly attend services and activities. The people of the town represent, according to the scientists at the New England Centenarian Study, the baseline lifespan for the rest of us, if only we ate well and took better care of ourselves – and, crucially, maintained social connections.

Surprisingly, there's a mini Blue Zone in central London. It's been there since 1682, and the people who work there say living there can add ten years to your life. It's a retirement home for soldiers from the British Army: the Royal Hospital Chelsea, home to the Chelsea Pensioners.

The hospital and grounds, designed by Christopher Wren, are grand and imperial; the place exudes wealth, like an Oxford college, its Hogwarts-style communal dining hall standing alongside modern hospital facilities. I have a coffee at the café when I arrive, and sit among veterans, some in the famous scarlet uniform with shiny brass buttons and medals, some in more casual fleece jackets, but still bearing the Royal Hospital Chelsea coat of arms. Despite the impressive Grade I- and II-listed architecture and its location in one of the most expensive parts of an insanely priced city, the hospital was designed for rank and file soldiers. You don't have to be rich to live here, you just need to have been a soldier. If you apply and are accepted, you sign over your military pension to the hospital, which is subsidised by the Ministry of Defence. And that's it. You then have all your needs taken care of. It's the security, but perhaps more than that, the community, that seems to be behind the longer lifespans here. I doubt it's the diet – you don't get Okinawan seaweed and miso-rich super-foods here. When I visited, in 2017, the menu included black pudding and fried pig's liver, devilled lamb kidneys on toast, and creamed semolina pudding. It was like seeing a menu from the war.

But then being here is like stepping back in time. I meet John Humphreys, ninety-eight, who shakes my hand with the grip of a man seventy years younger. It's no exaggeration to say he's lived the life of a Hollywood action hero. I wasn't sure he'd want to rake over old memories of the war, but he didn't mind telling the stories. They certainly reflect on his character. Gerontology researchers are finding that personality does seem to have an impact on lifespan, and with Mrs Love I was struck by her no-nonsense attitude. We sit in John's room, overlooking the beautiful gardens that run down towards the river, and he tells me about the time in 1942 when he was captured by the Germans in North Africa. He was twenty. Wounded and knocked out, he came round to see two large enemy soldiers standing over him. 'For you, Tommy, the war is over,' one said.

He was held in a POW camp in Italy. 'I didn't like being incarcerated in a blimmin' prisoner of war camp, so I decided to escape.' It took nine months before he had his opportunity. He spent the time learning Italian and used it to bluff his way past the guards dressed in a scrounged Greek army uniform that looked similar to the Italian uniform. Shepherding a couple of his mates in front of him, he told the guard in Italian that he was transferring them as prisoners to another location. The guard waved him through, and once clear of the camp John and his friends hiked south for days, living off the land. He recounts a heart-stopping moment when a German convoy passed by them – now dressed like Italian farm labourers – on the road. The final car of the convoy stopped and an SS officer got out. This is it, thought John: it was one thing bluffing his way past a bored Italian prison guard, but quite another to do it to a member of the fearsome SS. The Nazi officer called him over, and asked him, in broken Italian, where the nearest river was. John managed to explain, and the Germans drove on. John and his friends made it to a village held by British soldiers and got six weeks' home leave.

By 1944 he had joined the Parachute Squadron Royal Engineers, and jumped into Arnhem, for one of the major battles of the late stages of the war. Captured again, it didn't take him long to bust out. Some POWs he asked didn't want to try to escape as their morale was low, but he found three mates who were up for it. Again in the classic war-movie manner, he scraped the cement from around the bars of a window, replacing it with ash from the cooking stove so that the guards couldn't tell what he'd been doing. Eventually he'd done enough to remove the window bars and escape. The four men stole a boat and escaped down the Rhine to Nijmegen. On the wall in his room there's a photo of him and his fellow escapees in their boat. Humphreys looks every inch the invincible war hero. There's also a colourised wartime-era photo of his now-dead wife.

'I've always had a positive outlook on life,' he says; he thinks some people are just born that way. But he can't say whether his parents had the same attitude.

'I didn't know my parents very well,' he says, and I'm reminded of the generational gulf that separates us. 'My mother gave all her love and affection to my older brother. I was the gopher.'

The gopher?

'John go for this, John go for that.'

So much for nurture. Still, he seems to have been mentally and physically in good shape his whole life. 'I was in the parachute squadron, so you've got to be fit. My left patella was smashed.'

John describes how he got hold of a bike after he injured his knee, and carried out his own physiotherapy. Six months later he won the 100 metres and the 200 metres in the platoon athletics. 'You're not going to live a quiet life if you're a paratrooper. I've broken both collar bones, scapula, both wrists, right leg. But they healed fairly quickly. They don't give me any bother.'

He's clearly a determined and goal-driven man.

'My advice for a long life,' he says, 'is to be positive. Do what you can, the best way you can. If you have a setback, you've got to put up with that. You can't have a garden of roses without a few thorns.'

Nimmi Hutnik is a psychologist at London's South Bank University, where she specialises in mental health issues such as depression, anxiety, low self-esteem and PTSD. I came across her through a remarkable survey she and colleagues had made of centenarians living in the UK. Hutnik researches mental resilience (which we will explore more fully in the next chapter), and became interested in how it might be expressed in centenarians. Her team travelled around the country and interviewed sixteen centenarians, five men and eleven women. The researchers' approach was simple but effective: they started by asking the interviewees to tell their story, following up and prompting with questions such as: can you tell us something that has shaped your life? What is it like to live to a hundred and beyond? What's the secret of positive ageing?

The centenarians were self-selected. They had replied to recruitment notices for the interviews, so there may be an element of bias here, but nevertheless the general theme is fascinating. The centenarians don't seem to have achieved an extraordinary lifespan because they'd had an easy life, Hutnik says, but because they'd dealt with stress efficiently when it struck. They used phrases such as 'Accept whatever life brings', 'don't worry about the past', 'take each day as it comes' and 'do what you can to make things better and then forget it'. They described accepting things they could not change.[10]

Take Phyllis, who was 102 when interviewed. Quite a lot had happened in her life, she recounted: 'Different things, what with my father being killed and then my husband going to the war and my brother being killed and things like that. But then again, you're left and you've got to get on and that's it. You just have to cope with it, don't you? I'm afraid I'm one of those resilient people.'

And then there's Albert, also 102 at the time. 'If you can't change it and alter it, don't worry about it,' he said. Albert had been a miner since he was fourteen and the coal dust had damaged his lungs, but he still went dancing as often as he could. Nita, also 102, said: 'If you can manage and you've not got pain, you've got to try and push worry away and not be miserable.'

We've all probably heard from elderly relatives about how they just got on with things back in the day. The implication — sometimes it's more of an accusation — is that the current generation is too hesitant and coddled. That may be true, but then more of us are living to a great age, so we can't be that coddled. Perhaps it's just one of those things old people say and we'll say it when we're old, too.

When Mrs Love and I first start talking she mentions that during the war she had a baby who died. At first I don't pursue the subject, reluctant to press her on such a tragic event, even if it was something that happened seventy-five years ago. But later I ask her what her recipe is for positive living. Every life has its ups and downs, and even more so very long lives, so what's her advice on getting through it? 'You have to get on with it,' she says. 'I learned that when our baby died at one year old. It was devastating.' This last word is spoken with an exhalation, a sigh. 'But I thought: "You just get on with it."' You just had to pick yourself up and have another baby. 'There was no counselling in those days. You have to be positive to get on with things. No good everybody doing everything for you.'

For a while during the war this attitude was widespread, she says, but she thinks that not everyone has it naturally. This attitude, this positivity, is reminiscent of Ellen MacArthur and of Petra Kasperova, of Dave Henson and Barbara Hannigan. In fact it reminds me of almost everyone I've met while writing this book. 'You've got to have it in yourself to do it,' says Mrs Love. And she still has it. After she broke her hip in the summer, her daughter tells me, the physiotherapists were impressed with her mother's determination to get better. 'I still get out every day,' Mrs Love says. 'I was determined to get over that. Not my scene to flop around really.'

Mrs Love's comments – which I find humbling and inspiring at the same time – remind me of something I've read about Jeanne Calment. According to a biographer, she was biologically immune to stress. She had a saying, 'If you can't do anything about it, don't worry about it'. She also ascribed her longevity to her calm approach to stress, telling the zoologist and writer Desmond Morris, 'that's why they call me Calment.'[11]

When I first read this about Calment, I was immediately reminded of a common Japanese expression, *sho ga nai*. It means 'nothing can be done about it'. A variant, *shikata ga nai*, means something similar, or 'it cannot be helped'. After meeting Mrs Love, and reading what the centenarians in Hutnik's survey said to her, I am reminded of it again.

I noticed in Japan that Japanese people got less riled about things than I did. Sometimes it riled me that people didn't get riled. There are surely some things worth getting passionate about, I felt. The passivity and acceptance of fate encapsulated in *sho ga nai* got my goat at times. But eventually I found it admirable. The attitude certainly helps make a very pleasant and harmonious society to live in, and probably helps keep your blood pressure down. Whether it accounts for the long life of the Japanese is another matter. You can't *will* yourself to a longer life. We know stress has a range of important ill-effects on the body, but it's not enough to live calmly, or positively, or even to be determined not to flop around: we want to know what are the concrete links to longevity, and for that we need to explore the genome.

Neither John Humphreys nor Elizabeth Love have had their genome sequenced, but it's a fair bet that they carry a particular variant of a gene that has been the subject of huge amounts of research over the last twenty-five years. The gene is called *ApoE*, it lives on chromosome 19, and it makes a protein called apolipoprotein E. If you look at a picture of apolipoprotein in the protein data base you'll see a hefty, curly beast of a protein made of 299 amino acids. Its job is to transport cholesterol, and it comes in several forms.

Since 1994, when the gene was first linked with extreme lifespan,[12] it has spawned hundreds of research papers. From the number of people studying it and the amount of money going into it, you might think *ApoE* is the most exciting gene in the genome. But as with everything, it's complicated. One version of the gene, E4, has been linked to Alzheimer's and cardiovascular disease, and thus an early death, but an analysis of 2776 centenarians and a control group of younger people showed that if you carry either of two other versions, E2 and E3, you are more likely to reach extreme old age.[13] There is a longevity benefit to having E2, especially, and a cost to having E4.[14]

Kaare Christensen's influential Danish study of the role of genetics in ageing tracks cohorts of people born in 1895, 1905, 1910 and 1915. Christensen has found that the 'bad' *ApoE4* variant decreases in frequency with each successively older cohort, being found in about 20 per cent of the cohort at fifty years old and 10 per cent at a hundred:[15] 'If you look at centenarians there will be fewer with *ApoE4*. For sure, it's not a good gene to have but it's not like you're definitely weeded out if you have this gene.'

Christensen, a physician who works in public health, says one of the criticisms he's heard of his longevity findings is that modern society is failing because it helps weak people to scrape through to old age. This is called the 'failure of success': the success is that we're living longer but the failure is that we're in bad shape when we get there.

He set out to test this idea in a 2017 study.[16] 'We followed the 1905 cohort until they turned a hundred to see if they got more disabled, and they didn't.' He then looked at the 1915 cohort. Not only did more of them survive to ninety and a hundred, but they were in better physical and mental health than previous cohorts had been at that age. 'This is very encouraging,' Christensen says. 'It gives us reason to believe that in the next thirty, forty, fifty years we'll arrive at high ages with better cognition.' IQ has been steadily increasing throughout the twentieth century, in a phenomenon known as the Flynn effect, after its discoverer, James Flynn. 'This effect seems to hang in there when we arrive at high ages,' Christensen says. I'm reminded that Mrs Love does the cross-word every day.

Most of the variation in survival up to about the age of eighty is due to environmental influences: the diet you've eaten throughout your life, the support you've had from family and friends, the medical care you've received. Environmental components include behavioural factors such as whether you smoke, how much alcohol you drink, what exercise you do. But for the very old, genetics plays a much more important role. 'One reason is there's been longer time for your genes to make their marks on you in either a good or a bad way,' says Christensen.

However, despite the robust findings that *ApoE* contributes to longevity, the effects are seen only when you analyse large numbers of genomes. On an individual level, even a gene such as *ApoE* has only a small effect. 'Oh, it's trivial,' says Christensen. Its actual influence on our individual lifespan is tiny, just as any of the gene variants related to intelligence have only a tiny effect on their own. 'There's little doubt that there are thousands of genetic factors each with very small effects.'

To some this means there's no point chasing genetics to try and 'solve' ageing. To others, it just means the problem is harder, but it's no reason to quit.

Thomas Perls, based at Boston University School of Medicine, runs the New England Centenarian Study (NECS). The study was launched in 1995 to track centenarians living in the Boston area, primarily to investigate dementia. It has enrolled around more than 1600 centenarians and now has on its database around 150 supercentenarians, people aged 110 or over. It is the largest such sample in the world, and it has included the second oldest person ever to have lived, Sarah Knauss, who reached the age of 119. Knauss, incidentally, was another who seemed to handle stress well. She never let anything faze her, her daughter said. When Knauss was informed in 1998 that she was the oldest person in the world she replied with two words: 'So what?'

So what? The 'so what' is that she lived so long without major illness or cognitive decline. 'Supercentenarians,' says Perls, 'are the crème de la crème when it comes to compressing the time they are sick towards the end of their very long lives.'

Compression is a term you hear a lot from longevity researchers. 'Compressed morbidity' refers to the fact that the very old often stay healthy in body and mind right to the end, like the two elderly people we've met in this chapter.[17] This period of

213

physical and cognitive function is called the healthspan, and it is this that Christensen has found is growing longer in his Danish cohorts. It is here where we are most likely to see extension in the coming years, rather than lifespan itself.

The NECS team divide the subjects in their database into three groups: escapers, delayers and survivors. The first group account for about 15 per cent of the study members, and as their name suggests they have, remarkably, escaped any kind of serious disease. Elizabeth Love and John Humphreys are classic escapers (in John's case, literally). Around 43 per cent of the NECS members are delayers, and have avoided contracting any serious age-related disease until the age of at least eighty. Then there are the survivors, accounting for around 42 per cent of the members, who have had a serious disease before they were eighty, but have fought it off.

'Probably 75 per cent of their ability to get to their age is genetic,' says Perls. 'Thus they may be the key to our ability to discover longevity-associated genes and biological mechanisms that slow ageing and protect against age-related diseases.'

There are a few ways to measure this genetic component, a common one being, as we've seen, the genome-wide association study (GWAS). This is a way of scanning the genomes of many different people, looking for genetic variants that are associated with a disease or a particular condition – in this case, longevity. The method looks at the variants in the genome called single nucleotide polymorphisms, or SNPs. We each have many thousands of these tiny variations, and a GWAS attempts to identify those that are linked to the trait in question.

Perls's team have applied this to 801 centenarians, with 914 control people. Now, one of the criticisms of GWAS as a useful tool is that there may be thousands of SNPs relating to any given condition. That's certainly the case, as we saw, with complex traits such as intelligence. You need huge sample sizes to be confident about what you find, and 801 centenarians is not a lot. This means you can't seize upon a particular SNP and say, 'You need this to be a centenarian!' Still, it's a start, and a GWAS search represents an unbiased sift through a giant pond. Perls's team have found 281 SNPs that they call signatures of exceptional longevity.[18] Follow-up GWAS studies on a further 2070 people who were among the longest-lived 1 per cent of Americans born in 1900 identified yet more new variants associated with long life.[19]

It confirms, the team write, that exceptional longevity is influenced by the combined effects of a large number of SNPs. A 2014 study by Kaare Christensen and colleagues failed to find evidence for the same effect on longevity among those SNPs,[20] but according to Perls that's because the Danish study didn't look at extreme old age, only nonagenarians. Christensen is among those who say that to perform a decent GWAS scan you need far larger samples than is possible – there just aren't enough centenarians around.

All this has a crucial take-home message. It means you can do all you like in terms of diet (or dietary restriction) and pill-popping, but unless you've got the genes you just ain't gonna be a centenarian. Or as Nir Barzilai, director of the Institute for Aging Research at New York's Albert Einstein College of Medicine, put it to me: 'Environment may take you beyond eighty but not close to a hundred.'

It seems to kill the idea of Blue Zones, too. At least, it kills the idea that you can move there, or emulate the diet of the place, and reap the benefits. We saw from the work of Christensen that the environment doesn't seem to influence whether you reach a hundred or not, and Barzilai and colleagues have used a clever method to show how the environment, Blue Zone or not, is less important than we might think. The Albert Einstein College of Medicine runs the LonGenity project, a long-term study of the factors contributing to long life. The subjects are Ashkenazi Jewish, recruited from the north-eastern United States, which means they are from a similar genetic background, making it easier to spot any influential genetic factors.

The participants in the project, aged sixty-four to ninety-five, are divided into two groups. They are defined either as the offspring of parents with exceptional longevity (OPEL), meaning that they have at least one parent who lived to age ninety-five or older; or as the offspring of parents with usual survival (OPUS), where neither parent survived to the age of ninety-five. Barzilai's team test the participants on a range of physical parameters, such as balance, grip strength and mobility. They have found that people in the OPEL group perform better on average than those in the OPUS group: if your parent is exceptionally long-lived, you can look forward to a slower physical decline.[21]

Barzilai's team have gone further, and looked at the medical and lifestyle histories of the participants, including their diets. They have found no difference in calorific intake between OPEL and OPUS members, nor any differences in the types of food they ate – the proportion of fruit and vegetables, or grains, or meat. So the 'nutritional environment' for all members is the same. However, OPEL members have a 29 per cent lower chance of having hypertension, 65 per cent lower odds of having a stroke, and the risk of cardiovascular disease was 35 per cent lower, compared with OPUS members.[22]

'I think Blue Zones are mainly genetic islands and not environmental islands,' says Barzilai. 'I don't think we have capacity to survive over the age of a hundred with environment only.'

The sifting for genetic markers continues. In another study

by Thomas Perls and colleagues, two supercentenarians, a man and a woman, had complete genome sequences. The pair were selected from the NECS database because they had attained the age of more than 114, still without any disabling illness. In other words, as we've seen, the lifespan of these supercentenarian escapers closely matched their healthspan. When Perls's team examined the complete genome sequence of the two, they found they did *not* carry most of the known longevity variants. The two did, however, carry a number of variants associated with disease. They had a similar number of these disease variants to those found in regular, non-centenarian samples.[23]

So what does this mean? First, that the known longevity variants are not the whole set. There are many more remaining to be discovered. This idea is supported by the finding that the supercentenarian pair had many novel genetic variants at locations in the genome close to the longevity SNPs found in the previous study.

Second, it shows that you can live with 'disease genes'. Perls was surprised at first when the genome sequence showed disease-associated gene variants. Scientists had assumed that centenarians would have clean and tidy genomes, but they don't. Again, however, this points to the protective effect they get from elsewhere in the genome. This protective effect extends to ameliorating the damage done by smoking or drinking or having a poor diet. Remember, Jeanne Calment smoked for ninety-six years and Elizabeth Love had a seventy-year habit. The upshot is that there are hundreds of genes and thousands of variants of genes influencing longevity. It means we can't hope to tinker with them to make improvements, there are just too many. It's the same issue we saw with intelligence. Perhaps a better bet is to mimic their effects by using drugs.

*

It would be useful now to look briefly at attempts to live forever.

'Geriatric medicine can be defined as the attempt to elim-inate the diseases of old age directly,' says Aubrey de Grey. 'It has been an abject failure.' A prodigiously bearded real-ale enthusiast, De Grey is one of the central figures in the lifespan-extension movement. A longevity researcher, he is a co-founder of the SENS Foundation, a gerontology research institute based in California. SENS stands for Strategies for Engineered Negligible Senescence. The reason geriatric med-icine doesn't work, he says, is because the diseases of ageing are nothing like infections. 'They shouldn't really even be called diseases, because they are inseparable from ageing other than semantically. Being side-effects of being alive, they can't be cured.'

This is becoming a mainstream view among longevity researchers and public health officials. Ageing is the common factor in the diseases that hamper and kill us. Even a cure for cancer would increase the population of older people by only 0.8 per cent over fifty years, because other diseases – cardiovas-cular, diabetes, stroke – would take its place. Delaying ageing itself would increase the population by 7 per cent – and provide significant economic returns. Dana Goldman, professor of public policy and pharmaceutical economics at the University of Southern California, in Los Angeles, estimates that such a delay of ageing would generate $7.1 trillion over fifty years.[24] Jay Olshansky, at the University of Illinois, calls this the longevity dividend. We enjoy healthcare which means we have a relatively long life, but most of us suffer ill health at the end. This is very expensive, not to mention a poor way to end life. If we tackle ageing as a disease, we can increase healthspan and save money, and increase the vigour of the elderly.

Where de Grey differs from this kind of thinking is in

statements claiming that the first person to reach 1000 has already been born.

People will go to almost any lengths for a longer life. More than a hundred years ago, a French-American physiologist named Charles-Édouard Brown-Séquard reported that an injection prepared from the crushed testicles of guinea pigs and dogs could rejuvenate and prolong life. He was seventy-two. He also claimed that his own sexual prowess had been boosted after he ate extracts of monkey testes.[25] Even back then (it was 1890), and even in Paris, where they are more open to outré suggestions, his ideas were treated by fellow scientists with disdain – though, one suspects, with a twinge of envy in case it was true.

He was eccentric, for sure, but Brown-Séquard was not entirely wrong. He was among the first to show that substances in the blood have effects on organs in the body – substances we now call hormones. His elixir didn't do much for him, and he died at the age of seventy-six, but he has a whole range of spiritual descendants: people willing to try anything. Even, for example, having infusions of a child's blood.

In 2011 at Stanford University, Saul Villeda's team showed that old mice given the blood of young mice received a boost to the growth of their brain cells.[26] Then Amy Wagers at Harvard University and colleagues performed an experiment whereby an old mouse was physically joined to a young one, such that their blood circulation was shared. The old mouse was a 23-month-old with cardiac hypertrophy. That's when the heart muscle becomes thickened and the size of the chamber is decreased. It was joined to a healthy young pup, a two-month-old mouse with no heart problems. Four weeks later, the researchers found that the heart of the older mouse had returned almost to the size of that of the younger animal.[27]

Villeda is careful to state that his goal is to extend healthspan. It would be a bit much to publicly declare that he wants immortality through the blood of the young. He wants to compress morbidity and delay the onset of dementia. Wagers' lab and others are trying to isolate the factors in young blood that are responsible for the rejuvenation. One is a protein called GDF11, which young mice have much more of than old mice. When Wagers injected just the GDF11 protein into older, hypertrophic mice, she found that after a thirty-day treatment their hearts were similarly reduced in size. In 2015 Villeda, now at the University of California at San Francisco, found an 'anti-elixir' of life in the brains of old mice. The compound, a protein called beta2-microglobulin, operates in the immune system to help differentiate between the body's cells and invading cells, and is also active during the development of the brain.[28]

You can imagine the nightmarish scenarios that some people have dreamt up after hearing about this work. Or if you can't, here's Deborah Price, professor of social gerontology at the University of Manchester, and president of the British Society for Gerontology: 'In the extreme scenario you could get baby farms. Clandestine clinics where babies are sewn to old people.'

Gene therapy is another option. In 2016 an American businesswoman named Elizabeth Parrish made headlines when she announced that she had travelled to Colombia the year before to receive two gene therapy treatments aimed at prolonging life. Parrish, the boss of BioViva, a Seattle-based biotech company working to develop treatments to slow the ageing process, was forty-four at the time of the treatment.[29] The target is the caps on the ends of our chromosomes, called telomeres. These caps are made of around 1500 repeats of six genetic letters, TTAGGG. Each time the chromosome divides the cap shortens by one, until eventually it is unable to divide any more. At this point the cell has reached the end of its life. Centenarians have longer telomeres, and this is associated with protection from both diseases of ageing and cognitive decline.[30] The tantalising idea is that if you can extend the length of the telomere, you can stave off the ageing process. This has been demonstrated in mice, by a team of Spanish researchers in 2012,[31] although its effectiveness in humans is unknown.

Because BioViva had not done the pre-clinical safety work necessary to progress to studies on people, the US Food and Drug Administration did not authorise Parrish's experiment — hence her trip to an unnamed clinic in Colombia. She claimed that the therapy had reversed her cellular age by twenty years, but other scientists were sceptical both of the claim and of the effectiveness even if it turned out to be true. You can see similarities to Paris 120 years ago.

Other people starve themselves. Michael Rae is one. He works with Aubrey de Grey at the SENS Foundation. The recommended calorific intake for an adult male is 2500 calories; Rae, forty-six, has restricted himself to only 1900 calories for more than fifteen years. There are many more like him. For example, Dave Fisher, from Bracknell in the UK, is fifty-nine and has been on a punishing 1600-calorie diet for more than twenty-five years.[32]

Drastic diets extend lifespan in a range of lab animals, from nematodes to flies and mice, by up to 50 per cent. This works because on a restricted diet, stress-resisting genes are switched on, and it is through the action of these genes that protection from ageing is achieved. Mice on restricted diets are resistant to the usual age-related diseases such as cancer and heart disease. One concern for people trying this method of longevity enhancement is that calorie restriction does not seem to work in primates. In a 25-year study, scientists from the US National

Institute of Aging found that rhesus monkeys fed a third less than normal did not live any longer,[33] although the healthspan of calorie-restricted monkeys may be better.[34]

Is it worth it? To the half-starved men (and it does seem to be overwhelmingly men) who do this, clearly the answer is yes. They profess to love life so much they want to extend it beyond the usual allotted years; it's a shame they can't enjoy the intervening period a bit more too. De Grey said calorie restriction had looked mighty hopeful for a while, but no longer, although 'Lots and lots of experts in ageing who should be ashamed of themselves are still doing it, of course.' De Grey's preferred solution is to repair the molecular and cellular damage of ageing so it doesn't become a problem.

When de Grey came up with the concept of comprehensive damage repair as a mechanism to extend lifespan, he coined the term 'longevity escape velocity'. This is the moment when life expectancy starts to increase by *more* than a year, per year. This situation would mean we rejuvenate at a quicker rate than we age; we would outpace death. In some longevity circles it is the fabled moment they are aiming for; in others it is seen as irritating science fiction. Here is De Grey's current thinking as to when we might reach this point: 'I think we have a 50 per cent chance of getting there within about twenty years from now, so long as funding for the early-stage work that's going on right now improves soon. Otherwise it could be an extra ten years.'

Alex Zhavoronkov, CEO of InSilico Medicine, a biotech firm in Baltimore, Maryland, is more optimistic: 'I have no doubt that the escape velocity has already been achieved. The recent advances in deep learning and reinforcement learning, digital medicine, cancer immunology, gene therapy and regenerative medicine, made over the past five years give me confidence.'

Zhavoronkov points to the sheer number of scientists working in this area, and the progress he says is being made, especially in South Korea and China. We saw earlier that an increase in healthy lifespan could generate significant income for the world economy. I wanted to mention Zhavoronkov to show how despite calls for caution and moderation from many longevity researchers in academic institutions, the prize is such that dozens of other researchers are charging ahead. Zhavoronkov is an optimist, a futurist and a businessman. His main fear is that research isn't happening fast enough: 'We are likely to see a global economic collapse before the major developed nations embrace productive longevity as the new source of economic growth.'

I'm also keen to talk to Zhavoronkov because he has experimented with anti-ageing drugs in the past. In particular he's tested rapamycin, a drug that mimics the effects of the genes turned on during dieting, and which seems to offer the benefits of starvation without the pain.

Alan Green lives in Little Neck, New York City, off Interstate 495, which happens to be just a few miles past Flushing Meadows and the scene of Petra Kasperova's Sri Chinmoy triumph.

'When I was seventy-two years old, I had a 38-inch waistline and could not walk my dogs in the park without angina and shortness of breath going up small hills,' Green tells me. A physician, he trained at SUNY Downstate Medical Center in New York, qualifying as an MD in 1967. The angina forced him to confront the fact that he was growing old, and deteriorating. He started reading up on what was known about ageing, and came across the work of Mikhail Blagosklonny, at the time professor of oncology at the Roswell Park Cancer Institute in New York.

Blagosklonny's major interest is in rapamycin, which is used to suppress the immune system during kidney transplants, but which also extends lifespan in mice. The drug operates on a pathway called mTOR, the 'mammalian Target Of Rapamycin', which seems to be key to several important diseases of old age, such as dementia and cancer. Blagosklonny has admitted to using it experimentally in the past,[35] and Green decided to try it, using, he says, common sense to work out the dose.

'I am now seventy-four years old. This spring I just took up cycling. I cycled 1000K in May and then 1000 miles in June and plan to cycle 1000 miles in July and August. The reason for the cycling is just the incredible joy of feeling fit after suffering from ageing.'

He now has a 31-inch waist. 'My heart is good to go,' he says.

It's tantalising, and Green is a passionate, practically evangelical believer in the power of rapamycin, but the anti-ageing benefits of the drug are completely unverified in people. 'Rapamycin is not safe for humans yet,' says Barzilai. It's why he is trying to get the US Food and Drug Administration to approve clinical trials. He has one project ready to go, designed to test the diabetes drug metformin specifically for its anti-ageing properties (the trial is called TAME: Targeting Aging with Metformin). The drug has already been used for years for people with type II diabetes, so we know it is safe, but Barzilai wants to give it to people with cancer, heart disease and cognitive decline, to see if the drug alleviates their symptoms and prolongs their life. 'We need field trials,' he says.

But is ageing itself wrong?

The FDA and similar authorities are wary of admitting that

ageing is a disease in its own right. It would deny a fundamental aspect of our lives, in fact the very aspect that gives it meaning, some would say. Others simply don't care. Curing ageing is big business, a massive prize, and some of the biggest players in Silicon Valley are involved. Google launched Calico, the California Life Company, in 2013 to tackle the 'problem' of ageing. In 2014, Craig Venter set up Human Longevity Inc. to document the genome sequences of one million people. Its goal is to tackle the diseases of ageing and extend the healthy human lifespan. As mentioned earlier, the Methuselah Foundation aims, by 2030, to have produced technology where ninety-year olds can be as healthy as fifty-year olds. Aubrey de Grey's SENS Foundation is an offshoot of this. Whether it is wrong or not, millions of dollars and research hours are being poured into 'solving' ageing.

To me the moral issue around trying to cure a natural process is beside the point. It's not Gradgrindish to ask that we look at this scientifically. We try to cure cancer and heart disease and all the other diseases associated with old age and there's no dilemma there. But a better way to solve these problems – scientifically, technically, socially and economically – seems to be to slow ageing itself. The worry, of course, is that solutions will be available only to the rich.

I started this chapter by asserting that I wanted my days to be unlimited. But I can't know what I'll want when or if I reach old age. When our oldest daughter was born, we named her after her great-grandmother Molly, who was then ninety-three. In recent years the older Molly had lost her mobility and was pretty much confined to her house, where she still lived independently. I remember her saying that she had come to terms with her own mortality. It seems an extraordinary thing, to be prepared to stop existing. But at the same time it's understandable: many, if not all, your friends are dead, the lives led by your family, if you have any, are very different to yours. Add to that the ill-health that many elderly people face, and you can see why some people don't mind dying. It shouldn't feel taboo to say so. But then nor should it be taboo to try and escape death for as long as possible.

9

RESILIENCE

'The things you think are the disasters in your life are not
the disasters really. Almost anything can be turned around:
out of every ditch, a path, if you can only see it.'

Hilary Mantel, *Bring up the Bodies*

Many times, as I've been writing and researching this book,
I've felt humbled by the achievements of the people I've met.
That's the way it should be – it wouldn't be much of a book
on superhumans if I didn't feel impressed by these people's
abilities. But of everyone I've interviewed, one person stands
out, and that's Carmen Tarleton. Her story is at first deeply
shocking and disturbing, but bear with me, because for all its
exceptional brutality, Carmen's response – her resilience – is
even more extraordinary.

On 10 June 2007 Carmen, then thirty-eight, was at home
with her young daughters in Thetford, a small town in Vermont
in the north-eastern United States, when her estranged husband broke into the house. Herbert Rogers was looking for a
man he supposed she was seeing, but finding no man there, he
attacked Carmen. 'I just lost it,' he told police later. He beat
Carmen with a baseball bat so violently he broke her arm and

eye socket. Then he doused her in industrial-strength lye – a sodium hydroxide solution used in cleaning. One ear, her eyelids and much of her face was burned away. She suffered burns on 80 per cent of her body.

I met Bohdan Pomohac, one of her surgeons, at Boston's Brigham and Women's Hospital. 'In terms of injuries inflicted by another human being it's certainly one of the worst I've ever seen,' he told me. 'In brutality it was beyond anything we have seen.'

Carmen was first taken to Dartmouth-Hitchcock Medical Center in Lebanon, New Hampshire, the clinic where she herself worked as a registered nurse. She was then airlifted to Brigham, where she was put into a medically induced coma in an attempt to save her life. Her face was almost completely destroyed; her family were able to recognise it was Carmen only by her teeth.

When people experience terrible head injuries, doctors can put them into a barbiturate coma in order to shut down brain function. This prevents the brain from damaging itself further by trying to keep going without an adequate blood supply.

Carmen remained in a coma for three months while Pomohac and his team performed thirty-eight separate surgical operations. She was left covered in a patchwork of skin grafts, and these, together with all the blood transfusions she'd received over the months, meant that she had become immunologically vaccinated against at least 98 per cent of the population. She was blind, severely disfigured, and lacked many normal facial functions. She was in a lot of pain. But she was alive, and somehow, somewhere, the germ of something remained within her.

'Even when I first woke up from the coma I just knew it was such a big event, and it was so strange, that it had other meanings for me,' she says. 'I could help a lot of people.'

Carmen started doing inspirational speaking. 'I looked terrible and people felt so sorry for me, but I wanted to show people it didn't matter what I looked like.' It's the ultimate demonstration that it's what's on the inside that counts. Photos of Carmen from that time are shocking; it hardly seems possible that she could've lived, let alone coped. I think the most painful thing to see are her eyes. Because, with all the grafts, she didn't have proper eyelids, her eyeballs look out through small circular holes cut into the skin. She had synthetic corneas and couldn't blink, and the edges of the skin are red raw.

'It forced me, Rowan, to look at the big picture of what life is really about. And that's where I've had to go, because this horrific event occurred and I found a way through it. And not because I'm different or special, but because that's what was meant to be: to show people you can be involved in these incredible events and you can forgive, totally, and move on. And that's what I've been able to do.'

She says she has forgiven her now-ex husband for what he did. (In return for pleading guilty to maiming, he accepted a sentence of thirty to seventy years in prison. Seventy years – rather than life – was chosen as the maximum, as a life sentence triggers an automatic appeal, which no one, including the husband, wanted.) 'I was never religious,' she told me. 'I took responsibility for my life. I was not responsible for what happened, for what he chose to do, but I certainly was responsible from that day forward.'

Even after leaving hospital, Carmen was still in a lot of pain. The injuries to her skin and the multiple grafts caused tightness when they healed, and this caused all sorts of secondary problems with the mobility of her neck and spine. Carmen was unable to move freely without pain, and was taking huge amounts of narcotics to try and deal with that. Still she kept going.

Then, on Valentine's Day 2013, Carmen's life changed again. She became only the seventh American to undergo a complete face transplant. The operation was performed by Bohdan Pomohac.

Pomohac says that for a long time he didn't think of Carmen as a candidate for a face transplant, because she was so immunologically challenged. All those grafts and transfusions had primed her body to attack almost any donated tissue. But there were several pressing reasons why she needed a new face. The pain and the narcotics, yes, but as well as that, the opening onto her eyeball was getting bigger, and this was threatening the integrity of the synthetic cornea. Then there was the drooling, and the problems with speaking and eating.

The operation was a technical success, but then there was the immune system to deal with. Her body mounted a massive rejection of the face, and although for a period four weeks after the operation she was pumped up on numerous immunosuppressants, the face was still being rejected, and there was only one more drug left to use. A full dose would help suppress the rejection, but would also completely wipe out her immune system, to such an extent that any trivial infection would kill her. In the end, with Carmen's consent, they gave her a little of that final drug, and that turned it around. That last bit was enough to tame her immune system. She got better. Carmen says she felt that she had made a choice to live.

'It's possible there is a psychological force that carries people through but my official answer has to be that the statistics were still a little bit in our favour,' says Pomohac. 'Sometimes people can surprise you with their resilience.'

Carmen calls her face the gift of love. 'I think of my donor Cheryl almost daily. I have pictures of her, I have one of her scarfs on the handle to my closet.'

With most face transplants, the bone structure of the recipient is markedly different to that of the donor, so that when the face settles, the recipient doesn't end up looking like the donor. But with Carmen this isn't the case. Clearly she doesn't look exactly like her donor, but there is a definite resemblance. When she talked about what it was like to have someone else's face, it was the only time in our conversation where she faltered.

'That's a big part of my life, being grateful for the comfort I have in having, you know, her face,' she said. 'It's one of the biggest gifts I could receive.'

On 14 February, the fourth anniversary of her transplant, she was going out for dinner with her boyfriend, though on other Valentine's she'd done something a bit more unusual. 'Sometimes I see my donor's daughter for Valentine's Day.'

Carmen has struck up a friendship with Marinda Righter, the daughter of the woman whose face she now wears. It was Marinda who gave permission for surgeons to use her mother's face, after a massive stroke had rendered Cheryl brain dead. Marinda, who lives in Boston, described the moment she first saw Carmen, after the face transplant. She'd been warned that Carmen wouldn't look like her mother, but to Marinda, she did. They hugged, and Marinda touched her face. 'I fell in love with Carmen there,' she said, 'and I've never felt closer to my mom too.'

Carmen suffered physical injuries that could easily have killed her. That she survived is remarkable in itself, but what seems even more incredible is how she has not just survived mentally, but developed into something else — something that by her own account is better. It's that 'psychological force', which Pomohac is understandably reluctant to talk about, that I'm interested in.

Carmen told me she wouldn't go back and change what happened: she's grown too much. It's a pattern I saw in other people as I was researching this chapter, and it's something we'll look at again in Chapter 11. People suffer an incredible trauma, and find a way through it.

How do they do it? For physical trauma, it's not so mysterious that some people survive. As Pomohac says, statistically, some people will just make it through. When they do we're so surprised we fixate on it, and may even call it miraculous. We remember these people, extraordinary survivors like Arizona Congresswoman Gabrielle Giffords, who was shot in the head in 2011. We don't just shrug and say, well, it's not impossible to survive even being shot in the head. Statistics aside, medical care is good and the body is capable of amazing repair – perhaps better than we might give it credit for.

But it is the mental resilience that is sometimes more impressive. Carmen told me that we get too caught up in negativity. You've got to take control, and make your own choices. 'I don't live like the typical person on the inside at all,' she said. 'I have different beliefs that help me go when I'm going.'

Some people can ride out trauma and even prosper, when others suffer lingering physiological stress and fear – the hallmarks of post-traumatic stress disorder. While I was at Brigham and Women's Hospital I visited a psychiatrist to talk about this. David Wolfe is head of outpatient services at the evocatively named Building of Transformative Medicine. The funny thing, he told me, is that the people who suffer trauma but actually come out better aren't the people who are studied. If someone is coping well, they're sent home. Psychiatrists see the people who struggle. But perhaps, he said, people don't suffer as much as we think.

'Psychiatry has been as guilty of this as anyone. You assume that if something bad happens to someone that they're going to have some mental problem.'

We don't look at it the other way around. For example, many people may have had abusive situations in their childhood, yet most of them turn out fine. It's human nature, Wolfe said, to project – to assume that there must be something bad to come out of horrifying events. 'It happens when we see patients in hospital with very serious diseases. We think "they must be depressed – wouldn't you be?" but actually they're not.'

That's not to say, of course, that people don't suffer and don't need treatment. According to the United States National Center for PTSD, seven or eight out of every hundred people will experience PTSD at some point in their lives (and more women than men are affected). Some 24 million people in the US right now may have symptoms. What Wolfe is saying is that it's wrong to assume a traumatic event will be the same for everyone. More often than not it's more impressive how well people cope with trauma, which makes sense from an evolutionary point of view: it's advantageous to have hidden powers of resilience. 'The ability to power through adverse conditions should be there in our DNA,' Wolfe said. 'It's more of a puzzle why from an evolutionary standpoint we have problems with that sometimes.'

After a few minutes chatting with Alex Lewis, I honestly don't notice that his lips have been replaced by an oval patch of skin harvested from his shoulder. It's my Homer Simpson face, he says. The skin puts on fat faster than the rest of his face, so he can easily tell if he's getting fat, as his mouth gets chubby. It's certainly a distinctive and unusual look, not to mention that he is a quadruple amputee, but as I'm talking to him, I notice that while I'm looking in his eyes, my brain fills in his lower face with normal lips – it assumes he has normal lips. When I deliberately shift my glance and look directly at his mouth, the Homer face pops back. But generally I don't notice. I also don't notice because he's such an affable, normal bloke and it feels very natural chatting to him with a cup of tea in his kitchen. He drinks his with a straw.

In 2013 Alex, then thirty-three, was living with his partner, Lucy Townsend, and their young son Sam, running a pub in Hampshire. For an easy-going man who liked a drink, it wasn't the best situation.

'I'd be cleaning the lines in the pub and trying all the beers at ten in the morning. Then what do you do after that? People come to see you – you have a drink. Someone else drops in at two o'clock, someone at three – all the way through to kicking out at three in the morning. Every single day. It became the norm. But it shouldn't have been going on.'

It didn't really bother him, despite Lucy having a go at him about his horizontal ways. As he puts it now, he was in something of a rut – actually more a booze-soaked trough. 'I could easily put away twelve, fourteen pints and two bottles of wine, then wake up and do it again. I didn't feel like I was losing it.'

In November that year he had what he thought was a cold or a sore throat. It developed into something like flu, but then there was blood in his urine; a purple rash spread across his body. His fingers started to feel weird – he couldn't button his shirt. On the morning of 17 November, he says, he remembers Lucy coming back to the pub and banging on the door – it was locked and he had the key on the inside. As he came down the stairs he collapsed. 'I lost all function, all cognitive ability.'

As he lay there, completely out of it, Lucy and her father

prised open the door and called an ambulance – which was local, so it luckily arrived in five minutes. He was rushed to Winchester Hospital.

He was slipping in and out of consciousness that day as the medics tried to stabilise him. He had sepsis, an immune condition triggered by infections that causes the body to attack its own organs. But what was the underlying cause? One doctor thought he might have Weil's disease. Finally, at about ten that night, he was diagnosed correctly by a doctor who had seen the disease before.

The 'flu' was in fact an infection of the *Streptococcus* A bacterium. We all have these bacteria, living on our skin and in our bodies. If they get out of control they cause sore throats and pneumonia, but sometimes, rarely, they can develop into a serious infection called necrotising fasciitis. This being too medicalised a term for people to deal with, it's often called 'flesh-eating disease'.

This is what Alex had. By then his kidneys were failing and he was close to death. The team in the intensive care unit told Lucy and Alex's mother that if Alex didn't rally by the morning, they'd wake him for goodbyes, and take him off life support. His doctor that evening, consultant anaesthetist Geoff Watson, gave him a 3 per cent chance of surviving, but overnight he tried something unorthodox, and it saved his life. (Watson has never told Alex exactly what it was that he did, possibly because of its unorthodox nature.)

That was only the beginning. Things moved rapidly. Alex was moved to Salisbury Hospital, which has specialists in amputations and plastic surgery. The strep had penetrated deep, especially into Alex's left arm. Consultant surgeon Alexandra Crick came to see him. Alex greeted her. He said he was very polite, and he asked her how she was doing. The surgeon told him, 'Well, you're going to lose your left arm and maybe your feet,' turned around, and she was gone.

'I'm lying there thinking, "fucking hell, that was a bit harsh".' Soon, however, he grew to understand why Crick's manner is like that. Quite often people with his condition die, and doctors have to stay detached, to a certain extent, from their patients. It doesn't mean they don't care, but it means they can give the right care. 'And you can't dress it up,' Alex says. 'You have to give the patient accurate information, so they don't start dreaming up things that might happen.' He came to respect Crick's skill, and now he says there's no one he admires more than Alex Crick. He's had more than a hundred hours of surgery in three years. She says she will maintain a doctor–patient relationship with him for the rest of their lives.

Crick's initial, brutal assessment was correct. He did indeed lose his left arm – and both legs above the knee. 'It was such a quick thing,' Alex recalls. He began a long period of recuperation, and this is where his relaxed nature paid off.

'I just came out of it with a clear head and a clear idea that I was going to get better – not start again, but not be engulfed in grief and resentment and "why me" and all that.'

His outlook – remarkably similar to Carmen Tarleton's 'take responsibility for my life' attitude – was now a massive help to him. 'I was always a very easy-going guy. Even if I was drinking heavily, people would always say they could never tell when I was drunk. I never changed from twelve in the afternoon to two in the morning. I was placid; I didn't get stressed very easily.' What had kept him in a rut before, the go-with-the-flow complacency, now allowed him to let his limbs go without tearing himself up inside. And he found within himself a positivity and a strength that he didn't know was there.

There was no sign of this resilience in Alex's life before he became ill; he wasn't a particularly a tough child, nor did he show signs of inner strength before the strep. There's nothing he can recall. 'You don't really know what you've got within you until you plunge into the depths.' Carmen had said the same thing. She was a registered nurse and was raising kids, but until her incident she said she'd not thought about the big questions of life. 'I wasn't going to sit around and complain and cry.'

For Alex, obviously it's not all been an upwards trajectory since the strep hit him. The worst blip, if you can call it that, was losing his right arm. The surgeons had rebuilt it, and Alex showed me photos of the pioneering operation. The arm was butterflied open from wrist to shoulder, and the strep-ridden diseased flesh scooped out. It was filled in with muscle from his shoulder and stitched up again. But a few months later he rolled over on it in bed and it snapped. 'I sat up in bed and my hand was flopping down.'

The strep had gone into the bone – and the arm had to be amputated.

Despite everything – and this part of the story seems incredible, unbelievable – he feels thankful, overall, for what happened to him. He sees his life as changed for the better. You could be sceptical and insist, well, he would say that because what other choice does he have, but I believe him. Other people he's met during his treatment and rehab haven't been able to respond in that way. People like Alex and Carmen surprise you.

Alex says he and Lucy were very clear from the beginning that she wasn't going to be his carer. They now maintain lives that aren't solely centred on his disability. She runs a pub (The Greyhound on the Test, in Stockbridge, a different one from where Alex fell ill), Alex runs an interior design business, and they have the Alex Lewis Trust, which campaigns and fundraises for Alex's care and prosthetics, as well as for other amputee charities. Through the Trust he's skydived over Wiltshire and kayaked under the Northern Lights in Greenland. 'I live an incredibly satisfying life,' he says, 'born out of the strangest circumstance.'

He says the attitude he has now was hidden away somewhere. 'I didn't think I had it. When I came out of hospital, all of a sudden I'm doing things, accepting things.' His attitude now is that if he doesn't say yes to any opportunity that comes his way, he'll never know if he's going to enjoy it, or if it's something his son Sam might want to do in the future. 'We travelled the length and breadth of the country and abroad and we're where we are now because of a shift in attitude. I look at everything in a different way.'

He's firm that on balance what happened to him was positive. 'I think it is good. If I'd have carried on in the vein I was going I was at risk of losing my family – losing Lucy, being a single father, not seeing Sam. The prospect of that frightens me more than anything else. And I think the strep did me a favour to give me a bit of breathing space to get my head straight, then come out.'

No one does things like this alone. I'm not now talking about the amazing surgeons and medics who saved and rebuilt the lives of Carmen and Alex, but the friends and family who rally round. Your social network is key to how you deal with trauma. 'People who respond well tend to have positive relationships in their life and that translates into positive relationships with their treatment teams,' says David Wolfe.

When Alex was ill, word spread, and people started coming into the pub asking how he was. (All that drinking had paid off.) 'All of a sudden people were leaving 500 quid tips. It became obvious that I had this huge support network, through the pubs, all the people we'd met in the industry.'

The Alex Lewis Trust was born in order to do something structured with the large sums of money that were coming in.

Alex had Lucy, and his family, and his friends. His best friend Chris flew over regularly from his ski-instructor job in Courchevel, France, to help him. Chris has been a massive help. As the time drew close for Alex to leave hospital and move home, he became concerned about how it was going to work – in hospital there were staff, sometimes four at a time, to winch him in and out of his wheelchair when he needed the bathroom. How was he going to cope at home? His reconstructed mouth opening at the time was big enough to fit a 1p coin – how was he going to adapt things so he could live and eat at home? Simple things like cups and forks weren't right for Alex now. Chris said he'd come and live with him for the first six months. (Let's take a moment here to acknowledge Chris's awesome powers of friendship.) 'Without him being there I wouldn't have progressed as quickly as I did,' says Alex. 'It just meant we were adapting things as we went. Progression was always there.'

Similarly, one of Carmen's sisters moved to Boston so she could see her every day while she recovered, even before she came out of the coma. When Carmen left hospital, years before the face transplant, she went to her family. She saw a therapist for a while after she was attacked, but she says her experience was so far off the chart that nobody knew what to say to her. 'I cried at my sister's and my mother's and bitched a little, and after a year and a half I thought, that's not going to help me.'

Optimism, not surprisingly, is one of the character traits shared by people who respond well to trauma. 'The opposite is hopelessness, which is a feature of depression,' says Wolfe. 'Engagement, taking it on, taking on responsibility, and being active in the process – these things go a long way.'

When you ask people what keeps them going, Wolfe says, the number one answer is family and kids. People who are resilient look to the future – it's like we saw when looking at longevity, like we saw with Ellen MacArthur when considering focus. Successful people have goals, and work towards them. I'm reminded of Carmen again. She had a clear goal: 'I needed to find a way through this, because I wasn't going to go anywhere. I was raising children, there were things I wanted to do. My biggest motivation was I wanted to be a role model to my daughters.'

Until now we've only been talking about people's emergent traits, their personalities, their attitudes, their outlook on life. Let's now go deeper into biology, and look at what we know about the genetics of resilience.

Jason Bobe points to a poster on the wall of his office. We're in the department of genetics and genomics science at the Icahn School of Medicine, Mount Sinai Hospital, New York. It's high up on Lexington Avenue and sirens wail from the street below.

'One of the tendencies of being in a genomics department is to think the cause and the solution to every condition is genetics,' he says. The poster – a huge, detailed pie chart titled Determinants of Disease – is his way of addressing that tendency. A chunk of the pie, about a third, is taken by genetics. But then there is a bigger slice, some 40 per cent, taken by behaviour. There are slivers of various sizes for medical care, environment, social circumstance, and within each slice there are subslices and annotations picking out risk factors such as obesity, stress, nutrition and, where appropriate, known genetic factors. It's a complicated, detailed chart. The message is that genetics and biology are a major determinant of health, but there are many other reasons why people get sick.

'You could rename this Determinants of Resilience,' says Bobe. Our ability to be protected against insults to our health, he explains, could be due to biology, medical care, environment, behaviour or social circumstances. 'That's the big picture I want to situate this project in.'

That said, Bobe is part of a project that is squarely set up to identify people with some sort of genetic protection against disorders: 'We typically study disease by looking at the people who have the disease. But there is a whole group of people we're missing – people who have the serious risk factors for the disease, but don't get sick, or have only mild symptoms.'

The idea of the Resilience Project is to identify these people. If we find them – Bobe calls them genetic superheroes – studying them could lead to new therapies, or new methods to prevent disease.

There are some famous examples. Their stories are inevitably poignant, if not tragic, since they are often discovered in the midst of disease, surviving where others die.

Steve Crohn was a New Yorker who, as a gay man, lived through the full horrors of the emerging AIDS epidemic in the 1970s and 80s. He saw people die around him, years before the virus was identified. When it finally was, he assumed he'd also caught the disease; certainly he'd been exposed to it countless times. He cast around for a doctor to explain this, and finally wound up at the lab of a young virologist at the Aaron Diamond AIDS Research Centre, Rockefeller University. This was in 1994.

The virologist was Bill Paxton, now at the University of Liverpool, who had put a call out for gay men apparently immune to AIDS. He soon found that Crohn's cells were immune to HIV. Paxton exposed the cells to three thousand times the amount of HIV that usually causes infection, but the cells shrugged it off. Quite literally, it turns out – HIV enters specialised white blood cells called CD4 cells via a molecule called CCR5 on the surface of the cell. Crohn had a mutation which meant he lacked CCR5, so the virus simply couldn't get a purchase on the cell. Crohn's variant – now known as the delta 32 mutation – led to the development of maraviroc, a drug which blocks the CCR5 receptor. Understanding what Crohn's variant did has also helped in developing strategies to cure HIV. Crohn, however, killed himself in 2013.[1] He was immune to HIV, but not to its tragedy, the *LA Times* noted.

Then there's Doug Whitney, from Port Orchard, Washington. Doug's mother developed early-onset Alzheimer's, a form of the disease caused by a single genetic variant, when she was fifty. As is sadly all too often the case, she died not long after. Nine of his mother's siblings also died, and the disease killed Doug's older brother at fifty-eight. Doug, naturally, braced himself for the onset. But it didn't happen. It still hasn't – he's now sixty-eight, way past the age this form of dementia normally shows itself.

Thinking he'd dodged the genetic bullet, a few years ago Doug had himself tested, only to find he did in fact have the gene for familial Alzheimer's. It can only mean that something else in his life or in his genome is protecting him from what otherwise is inescapable.

Bobe calls these cases 'inflated airbags'. It's the opposite of a smoking gun – if you're trying to solve a crime, the presence of a such a weapon is evidence that a bullet has been fired. In genetic medicine the smoking gun is like finding you have a gene for a major disease. But in resilience they're looking for the inflated airbag – a protective factor that has been deployed and has saved you from disease. In Steve Crohn's case the airbag was the delta 32 mutation.

Whitney, too, has an airbag, but we haven't identified it yet. And herein lies the scientific challenge. To understand why, let's look at a paper Bobe published in 2016, in collaboration with some thirty scientists from around the world.[2]

The paper, in the journal *Nature Biotechnology*, is an analysis of genetic data from more than half a million people. The collaborative approach allowed the scientists to share genetic data from twelve different studies. This is important because it allowed them access to samples from 589,306 healthy people. Healthy people aren't usually the ones who turn up in genetics studies.

The scientists then combed through the data, focusing on 188 areas in the genome where there are known disease-causing mutations. They were interested in what are called Mendelian diseases: diseases that are caused by a single damaged gene. Cystic fibrosis is an example. The team identified 15,597 people who carried a mutation. Remember, these are samples from healthy people. It's like what Paxton did when he put out a call for people apparently immune to HIV, except it draws on the greater analytical power of genomics.

These nearly 16,000 people were then assessed more rigorously to ensure the genetic data were reliable, that they really did carry a disease gene, and that they really were symptom free. This whittled the number down to just thirteen people. Thirteen unexpected heroes with a secret power – protection from otherwise severe genetic disease.

The thirteen have dangerous mutations – for diseases such as cystic fibrosis, Smith-Lemli-Opitz syndrome, which causes learning difficulties, and a disorder called Pfeiffer syndrome. This latter is a condition in babies where the bones in the skull fuse. The brain grows inside the skull but the bones can't grow to accommodate it. (I've seen a baby with this terrible syndrome, and watched the operation to relieve the pressure in his head. As you might imagine, it involves sawing gaps in the fused skull. The surgeon told me the pressure becomes so great inside the skull that the poor child's eyes can pop out of their sockets.[3]) These are gruesome and often fatal diseases, but something in the thousands of other genes these thirteen people carry is protecting them.

We visualise genes as beads on a string – or I do, at least. But that massively underplays how complicated they are, and how long the string is. Genes are more like trains on a railway track. To find the protective genes using Bobe's approach – to find the airbags – will require even larger data sets, which is why I mentioned the collaborative approach they are adopting. They'll need even larger samples to find more unexpected heroes and to find how they're protected. This is what they're doing – the Resilience Project, run out of Mount Sinai in collaboration with Sage Bionetworks based in Seattle, Washington, is aiming to get sequence information from one million healthy people. They'll then comb through the vast amount of data looking for the tiny changes that might protect certain individuals.

There is another way. Instead of searching for rare genes in a diverse population using huge numbers of people – Bobe's approach – you could look for them in much smaller groups of genetically similar people.

You know the places where everyone knows everyone and going there is like stepping back in time? Think of somewhere like that, then make it even more remote in time and space. Say, an isolated mountain village on the island of Crete, or a group of Old Order Amish in Pennsylvania, or an Inuit village in Newfoundland. These places don't tend to have a lot of new genes coming in from outside. It means there is less genetic diversity in the group, so rare genes may be present at a higher frequency than normal, and therefore easier to find. It's why the music-gene hunters we met in Chapter 6 went to Mongolia.

This has been the approach of Eleftheria Zeggini, of the Wellcome Trust Sanger Institute just outside Cambridge, UK. She studies isolated populations in Crete, Pennsylvania and Newfoundland.

'The villagers in Crete have lamb for breakfast, lunch and dinner,' says Zeggini. Their diet is incredibly high in animal fat, and sure enough, the people there become obese and get type II diabetes at the same rate as the general Greek population. But they don't suffer the medical complications usually associated with diabetes. The villagers instead are known for their good health and long lives. Suspecting there might be protective genetic factors at play – inflated airbags, in Bobe's phrasing – Zeggini moved in and collected as much information as she could from the villagers, including medical information such as blood pressure, blood samples to measure fat content, questionnaires on diet, and DNA samples.

In a telling illustration of how quickly genomics has moved in the fifteen or so years since the first laborious and expensive sequencing of the human genome, Zeggini's team simply went ahead and sequenced the total DNA content of more than 1500 villagers. (Bear in mind that the original human genome project took a decade and $3 billion to complete – at the Sanger Institute they can process a complete genome in about thirty minutes.)

By analysing all the information Zeggini found that the villagers tended to have three rare gene variants that are either absent from the general Greek or European population, or present at a much lower frequency. These variants are associated with cardioprotection: people with the variants process fat more efficiently than normal. 'That's interesting because we know their diet is bad, but that they don't die at the rate you'd expect,' she says.

What's also funny – in a good way – is that one of these variants popped up in the Amish population Zeggini studies. They too have a diet that is high in animal fat. If she had done the study on the cosmopolitan British population, with all its mass of genetic diversity, she would have needed to sequence 70,000 people to get the statistical power necessary to find the variants.

Scientists such as Bobe and Zeggini are at the forefront of the search for the factors that cause and protect against disease. Note that I said factors, not genes. It's not just genes. As Bobe told me in his office, there are many reasons people get sick. Even when we find genes related to disease, it rarely works on the basis that 'you have the gene, you'll get the disease'. Sure, there are the Mendelian diseases mentioned above, such as familial Alzheimer's or cystic fibrosis, where you almost always do get the disease if you have the gene, but most diseases, and almost all traits, are far more complex, and are influenced by many genes. Resilience itself being one such complex trait.

'How can I make myself and my family more resilient to disease? That's what I'm trying to do,' Bobe says. 'We're trying to systematically study that.'

Some of these things you can teach. Ann Masten, a psychologist at the Center for Neurobehavioral Development at the University of Minnesota, Minneapolis, calls the power of resilience 'ordinary magic'.[4] It is magic that anyone can use. John Humphreys, the war hero we met in the last chapter, said he thought he was born with a positive character, but this sort of thing can make people uncomfortable. It's not a big step from that to blaming people who aren't relentlessly positive for their outlook on life. But far from it. Nimmi Hutnik's team at London's South Bank University say resilience – although a complex mix of biology, psychology and environment – has the potential to be taught. Pharmaceutical interventions to extend healthspan are being developed, but until then, it's worth noting that exercises in mental resilience can be learned, and can be used to promote health and well-being.

The capacity to be super-resilient may be there even in us normal people, but we need guidance and support to find it, maybe from psychotherapy, maybe from friends. We need help to be optimistic, encouragement to take control, and empowerment to be responsible. We need a certain amount of self-love. A touch of narcissism is good! We need to stand up for ourselves so we are not mistreated at work or in relationships, we need to be assertive without devaluing others and have a self-image that is positive without being conceited. This mixture of personality traits will drive you forward. Some of them can be constructed, if you do not have them naturally.

10

SLEEPING

I lie about to fall asleep . . .
In the slot between waking and sleep
a large letter tries to get in without quite succeeding.

Tomas Tranströmer, *Nocturne*

Sleep that knits up the ravelled sleave of care
The death of each day's life, sore labour's bath
Balm of hurt minds, great nature's second course,
Chief nourisher in life's feast.

William Shakespeare, *Macbeth*

I'm having my head measured and divided into quadrants, and my scalp marked with a red pencil: I'm like a side of beef being marked up by a butcher. The marks will be the locations for the electrodes that will record my brain activity overnight. David Morgan, the Ph.D. student who will be monitoring my sleep from a control room up the corridor, scrubs at points on my scalp with an earbud dipped in abrasive gel, to clean the skin of dead cells and tenderise it for the electrode. 'Do behind the ears really well,' advises Morgan's supervisor, Jakke Tamminen. 'There will be a *lot* of bacteria there.' To me he says, 'No offence, most people don't wash behind their ears.'

I end up with an electrode grid that will pick up signals from the frontal, temporal and parietal lobes of my brain; as well as next to my eyes, to measure their roving movement when I'm in REM (rapid eye movement sleep – see below); and my jaw, to measure muscle tone: during REM sleep the body is paralysed, to stop you acting out your dreams. A final electrode is stuck Hindu-like in the centre of my brow, like a *bindi*; for all I know it's the chakra site of my third eye. The wires are gathered in a kind of cyborg ponytail, tied behind me, and plugged into a wall-mounted unit that will receive and amplify the signals, my brain waves.

We say 'I've had a brainwave' to describe the sudden genesis of an idea, but brain waves really exist, and come in different sizes, just like the waves in the ocean. They are the electrical activity of neurons in the brain, and when we are asleep they synchronise enough to make discernible patterns. I'm in a sleep lab in the department of psychology at Royal Holloway, University of London, to find out what mine look like.

Sleep is something of an anomaly in this book. It is perhaps the least well understood of all of the traits I'm investigating, having been scientifically studied for only a relatively short time. What is it for, exactly? We know that cellular repair and maintenance go on during sleep, and this is part of its restorative function. It also plays a role in memory storage. But *how* does it do this?

It's also odd in this book because it's not clear, at first glance, what it means to be a good, let alone superhuman, sleeper. Is it someone who sleeps only five hours a night and still feels great? Or is it someone who sleeps more than ten hours but is at the peak of their profession? We'll look at examples of both types, and we'll also meet someone who thrives by customising her sleep and breaking it into chunks sprinkled throughout each twenty-four-hour period. We'll meet people, too, who can control one of the most celebrated aspects of sleep: dreaming. Despite the uncertainty I've included sleep as a category because, quite simply, it is universal – not just to humans, but to all life forms that we know of – and because ultimately it is vital that we sleep well. Happily, unlike some of the other traits, sleep is something we can all excel at.

When Tamminen is satisfied that the electrodes stuck to my head are recording properly, they bid me goodnight and disappear into the control room. It's around 11 p.m., and through the intercom they say they will wake me at 7 a.m. the next day. Now, they tell me, in a similar manner to how I implore my daughter, please go to sleep.

Sometimes incidents in life lead you down peculiar and unexpected paths. In 1892 Hans Berger, a trainee soldier in the cavalry of the German Empire, was thrown from his horse into the path of a horse-drawn cannon. You can imagine that heart-stopping feeling, when time seems to slow down in the face of approaching death. In this instance it was in the guise of stomping horses dragging artillery. But death didn't come – the cannon stopped in time and Berger survived. Here's a weird coincidence: many miles away his sister had an intense feeling that her brother Hans was in danger. She begged her father to send Hans a telegram, checking on his safety, and when Hans received it, he was understandably stunned. How could she have known? He became obsessed with discovering how his brain could have possibly transmitted a signal to his sister.

Berger became a psychiatrist. In his career he was driven all the time by the desire to understand the energy of the brain, and in 1924 he recorded the first human electroencephalogram (EEG). EEGs have been somewhat overshadowed by MRI scanners, which can show at a high resolution what's going on deep in the brain, but the advantage of EEG (aside from its much lower cost) is that it allows researchers to watch changes in brain activity each millisecond, something not currently possible with MRI. It's also much easier to sleep overnight with EEG than in the growling, immobilising coffin of an MRI scanner.

And indeed I slept quite well in the sleep lab. The electrodes all over my scalp and face didn't really bother me – it was only the wailing of drunken students outside, and an unfamiliar pillow, that disrupted my slumber. I had a few peculiar dreams that I remembered after waking from some student singalongs in the middle of the night. It hadn't bothered me that there were people monitoring me as I slept, but in the morning, when I was shown my night's sleep on the computer, I realised how intimate an insight the scientists had into me. 'You fell asleep quite quickly,' says Tamminen, scrolling through the EEG traces showing my brain activity. 'You are drifting off here. There's a disturbance here, but that's just you moving around for some reason.' I met the man only yesterday, and he already knows how I sleep.

There are five stages of sleep, numbered 1 to 4; the fifth is REM sleep. Throughout the night we cycle through the stages and start again at stage 1. The proportion of time we spend in each stage varies throughout the night, and for any number of other reasons, such as our age and whether we're intoxicated or on medication or disturbed by something. Stage 1 is light sleep, and the transitional stage between waking and sleep. Usually I pass quickly from this stage, but in the sleep lab I was tossing and turning as noises from outside dragged me back to the surface. Tamminen shows me the long tail of oscillations before more interesting things start happening. What he's looking for is a spindle. Spindles are the sign that I've gone into stage 2.

In the mass of jumps and oscillations that make up a night's sleep as recorded on an EEG, you see odd, particular bursts of activity lasting half a second or so, always in the frequency range of 12–14 hertz. These are spindles, and originate from some brain spasm in the thalamus, deep in the centre of the brain. Spindles might have something to do with the integration of new information into the brain, because they seem to make the brain more plastic, that is, more agreeable to the acceptance of new data.

Tamminen explains that as we learn new things in the day, they are quickly encoded by the hippocampus. This acts as a short-term memory store, but at night the information needs to be consolidated. It is conveyed to the neocortex, a large and important region of the brain that looks after language acquisition and sensory perception; a region, basically, that is highly active when we are thinking. Spindles may act as gatekeepers to this region. There is some suggestion that people with spindles of slow duration are more intelligent,[1] so I am a little crestfallen when Tamminen says my spindle isn't a very good one.

We carry on scrolling through the electrical output of my unconscious mind. It's amazing what Tamminen can read into what to me are chaotic spikes and scratches. We find a nicer spindle. My brain waves, he says, are starting to slow. These are delta waves, and they first appear in stage 3 sleep. We carry on, into stage 4, and the waves become more prominent. This is slow wave sleep, and stages 3 and 4 are together called deep sleep. The waves are now gently rolling, and even my untrained eye can see the undulations. Here I am dead to the world. It's hard to rouse someone from deep sleep, and if you manage it, they are often disoriented and bleary. It's like they've come back from a strange land – looking at this readout, a strange ocean would be more apt – and the reason is they are bleary is because they are rebooting their operating system. It takes a while to get back online.

We continue scrolling, paddling, through the waves of my night. Soon the EEG starts to change again, and Tamminen says I am entering REM sleep. The electrodes wired to my eyes start showing peaks and troughs, indicating that my eyes are rolling around under my lids. I always wondered, are the eyes just unmoored during this time, or are they looking towards things you see in dreams? The answer is probably the former, as dreams can occur, we now know, in non-REM sleep, and it's not clear that we're always dreaming when we are in the roiling stage of REM sleep. The EEG wired to my jaw goes flat, indicating that my body is now paralysed.

Deep sleep tends to happen more in the first half of the night, and periods of REM sleep get longer in the second half. Sleepers tend to wake momentarily after cycles of REM sleep, and Tamminen, the Morpheus of this sleep lab, points out a sudden burst of activity on my EEG.

'You're awake here.' He checks the time stamp. 'It's about five in the morning. Do you remember this?'

I do, actually. I'd woken up and could hear blackbirds singing outside, and thought it might be time to get up. I'd been dreaming of sailing on a ferry to Liverpool, and seeing the Liver Birds on the Royal Liver Building at the waterfront. I soon went back to sleep.

'You're getting restless now,' Tamminen the Sandman says, as we approach the end of the night and the ocean gets choppy. 'You'll wake up soon.'

*

254

Back at home after my night in the sleep lab, I'm staring out the window, watching a fox curled up on the grass in the garden, dozing in the sun. His ears twitch as car doors bang in the nearby street, and he looks up when children shout in the neighbouring gardens. He's never going to go into deep sleep, is he? He can never relax. I look out again a bit later and the fox is still asleep, but has shifted position under the apple tree so as to avoid the shade and track the warmth of the sun.

Foxes sleep in snatched periods of time across a twenty-four-hour cycle: their sleep is polyphasic, as opposed to our monophasic sleep. Many mammals sleep like this, and especially those with small body masses, because they burn energy so quickly, even while sleeping, that they need to wake up and forage. Dogs and cats are similar, and it was the example of such animals, along with the prolific number of flights he was taking and the sheer amount of stuff he wanted to do, that inspired the American inventor and architect Buckminster Fuller to adopt a polyphasic sleep pattern.

Fuller, who died in 1983 aged eighty-seven, was a man who marched to his own drumbeat. When he was in his thirties, Fuller would work, eat and live – but mainly work – in four six-hour blocks, taking a thirty-minute nap in between. He called this hellish schedule Dymaxion sleep, and supposedly kept it up for two years, but even when sleeping like a regular person he was known for being tireless, driven and productive. There's no doubt that he achieved a huge amount in his life – he published more than thirty books, and among his many inventions, the geodesic dome is perhaps his most well known. Indeed, when a sleep researcher mentioned his name my first thought went to the football-shaped sixty-atom carbon molecules named buckminsterfullerene because they resemble his geodesic domes. As well as his inventions, Fuller was inspirational and forward-thinking, coining the term 'Spaceship Earth' to capture the idea that we are all living together on the planet, and that we must use our energy and resources renewably.

Fuller was undoubtedly a superhuman sleeper. Few of us will contribute to humanity as much as he did. How did he do it? Did he force that sleep pattern on himself, through strength of will? Or was there an inner drive that enabled him to work, that sourced him with energy despite the lack of sleep – was it, indeed, some inner energy that *stopped* him sleeping for very long?

Even if our goals are more modest, we'd like to contribute more to society, to ourselves and our friends and families; we'd like to have Fuller's tirelessness. We want that extra time. His example has helped inspire a community of sleep hackers, who insist that the eight-hour-a-night standard is unnatural, or is holding us back. Or at the very least they conclude that it is not for them.

Fuller is long gone to the eternal sleep, but modern pioneers of polyphasic sleeping carry his torch. Marie Staver, now a project manager in Boston, adopted what she named the Uberman system when she was a student, struggling with essays and revision and constant tiredness. She was also troubled by insomnia. Inspired by Fuller, she decided that the regular monophasic way of sleep just wasn't for her. If 'Uberman', by the way, sounds a bit like 'Übermensch', that's because the latter is what Staver and a friend originally called it. Perhaps they were trying to reclaim the word from association with Hitler. In any case, they eventually settled on the marginally less Nietzschean 'Uberman'. The system entails sleeping in six 20-minute naps, one every four hours. You get a total of two hours' sleep in each twenty-four-hour cycle – a superhuman twenty-two hours' waking time altogether. Staver says that the initial adjustment to this radical sleep pattern is monstrous, with symptoms of flu, headaches, and bouts of anxiety and depression. But once you are used to it, after a couple of weeks, it pays off because you become much more productive *and* you feel more rested. 'It took some work to maintain, but considering the benefits, it was definitely worth it.' You have to set an alarm to get you up after the twenty-minute nap, but she says she'd wake up at nineteen minutes, on the nose.

Staver is no longer an Übermensch – she could keep that up only for six months before what she calls 'social factors' got in the way. But for the last nine years she's been on a differently patterned polyphasic system called Everyman 3. Under this regimen, you get three hours' sleep in a block at night, and then three 20-minute naps throughout the day. It gives you four more hours of waking time than the rest of us in each twenty-four-hour cycle.

Staver says it's hard to say whether Everyman is easier than Uberman. Perhaps it's like asking an ultrarunner if a twenty-four-hour track race is easier than a 100-mile mountain run. 'For most people, Everyman 3 fits the most easily into a regular schedule, so it requires a less jarring lifestyle modification [than Uberman].' If you are on Everyman 3, Staver says, you can work a nine-to five-job, as long as you can get a nap at lunch. Because it has a core three-hour sleep, it involves milder sleep deprivation at first, but takes longer to adapt to overall. 'For me, right now E3 is easier because I still don't have the kind of lifestyle in other ways that would support a Dymaxion schedule – but I'm working on it, and the second I do, I'm switching,' she says. 'I love E3 compared to monophasic. I have four additional hours per day *and* I feel more rested.'

I do get the sense that she feels E3 is cheating, somehow. 'I love the pure polyphasic schedules best,' she admits. 'I suppose they are more challenging, but aren't all awesome things?'

As I have with many of the people I've met in this book,

I feel I am a long way from Staver's ability. I'm a boring old monophasic sleeper. People talk about 'pulling all-nighters' but I remember doing this properly only once, when I was writing up my Ph.D. I worked through the night in the lab, and met my friends at the regular departmental coffee break the next morning, wearing the same clothes. Then I staggered home and flatlined for hours. I can't imagine how much effort it is to keep to a polyphasic sleep pattern, but Staver says that she found it to be a lot of effort just to maintain monophasic sleep.

She can easily sleep in odd locations such as cars, on spare couches, even outside. She does experience the sensation of falling asleep as soon as her head hits the pillow when she is adapting to a new schedule, but says that once she's adapted she drops off in around five minutes. A five-minute sleep latency, as they call the time it takes you to fall asleep, is just about healthy and normal, but five minutes out of a twenty-minute nap time is a significant lost chunk. Does she *really* feel awake and alert on this schedule? 'I wake rested, and don't usually experience any tiredness until it's almost the next nap time. I certainly don't feel like I really need sleep any more often, in fact much less often, than I did while I was monophasic.' It's hard to judge this independently, as we don't meet, and only communicate by email.

I do wonder about loneliness.

'It took a little getting used to, being awake for so much time that others were sleeping,' Staver says. 'But it's very productive time, and if I need human interaction it's easy enough to find things to do with people during weird hours.'

She calls her extra time non-people hours, and says she mostly uses them for writing, and for taiji practice. Taiji is the Chinese Taoist meditative practice which is the basis of the martial art tai chi. Taiji means something like 'supreme extreme' or 'great ultimate' and refers to a state of infinite potential; maybe I'm primed to make the connection but it sounds to me quite similar in tone to Nietzsche's Übermensch.

Staver is quite an evangelist for polyphasic sleep. While of course she doesn't advocate it for everyone, neither does she think monophasic sleep is for everyone. There is a large and enthusiastic internet community of polyphasic diehards. And many people clearly experience a chronic lack of sleep. A US survey between 2004 and 2007 of 66,000 civilian workers found that 30 per cent reported sleeping six hours or less per night. For people in senior management, the figure was 40 per cent. Staver feels that the answer to this ever-increasing societal pressure to work is not to stop bringing smartphones into the bedroom, or to educate people on the need and the benefits of at least seven hours of uninterrupted sleep, but to look at other options to hack the system: 'Being long-term sleep deprived or constantly desynchronised is terribly unhealthy. We know this, and yet as a society we do *very* little, or nothing, to help encourage people to get good restful sleep, or to give them options when "be unconscious for eight-plus hours" isn't a viable one.'

It's almost as if there's a monophasic tyranny forcing the majority of us to sleep in eight-hour blocks. There are all these people who need more sleep, and we're putting more and more pressure on each other to be awake and productive all the time.

She's right of course that it is unhealthy to be sleep deprived. And we definitely all need more sleep. But my biggest concern about polyphasic sleeping is the health consequences. Doesn't it take a toll on you?

'The adaptation process does take a toll on one's system, and I certainly don't recommend optimising sleep schedules to the sick or frail,' Staver says. 'But after you're adapted, nothing I've seen makes me think that being polyphasic is doing any damage. Both Bucky Fuller and I have seen doctors regularly while and after being polyphasic, and were in excellent health.'

Polyphasic sleep is studied professionally. NASA knows that astronauts tend to manage only six hours' sleep per night,[2] and experiments have showed that naps can help boost working memory. The yachtswoman Ellen MacArthur had to adopt a polyphasic sleep pattern when she sailed round the world. She was advised by neurologist Claudio Stampi, who runs the Chronobiology Research Institute in Newton, Massachusetts. Stampi specialises in helping solo sailors cope without sleep.

There was a key difference between MacArthur's sleep during that time and the precise timing of Staver's system. MacArthur grabbed sleep when she could, but often had to leap up to see to something. 'The biggest thing that affects your sleep is the amount of adrenalin that is pumping through your body,' MacArthur told me. Her journals from the record attempt are full of entries about how her veins are so full of adrenalin she can't sleep, her mind is whirring, but she is exhausted. 'At any point during the round-the-world, with a few exceptions, you could capsize that boat any time, and you sleep with the ropes in your hands,' she said. 'So when you sleep it's five minutes, nine minutes, occasionally you sleep for twenty minutes, very rarely one hour.'

When I met her, she was clear about the single hardest thing about sailing around the round the world on her own: 'It's not that it's not physically hard, but it's the sleep deprivation. Sometimes you can't sleep at all. It is too dangerous.'

It's one thing to sleep polyphasically for the duration of a space mission, or if you are a fighter pilot, or even if you are sailing round the world on your own. It's quite another to try and do it as the norm, and the fact is that we just don't know what it might do in the long term, because we don't know enough about what sleep does. Time to visit some more sleep scientists.

To find the Guy's Hospital Sleep Disorders Centre you have to turn down an easily missed alley opposite Borough Market in London. I hear someone describe it as Diagon Alley. I meet Guy Leschziner, consultant neurologist and clinical lead for sleep. Also present is Meir Kryger, a legendary sleep scientist at the Yale School of Medicine who happens to be based at Guy's Hospital when I visit. Kryger was the first to diagnose sleep apnoea in North America, and has published many books and hundreds of papers about sleep. For good measure we are joined by Adrian Williams, a consultant sleep physician at Guy's and founder member of the British Sleep Foundation. 'If you want to find out what sleep does,' he intones, 'you deprive animals of sleep.'

At the turn of the last century the Russians did this with dogs. They found that the dogs died within three days; if deprived of water, they lasted eight or nine days. 'So sleep is more important than water,' Williams says. He delivers this remarkable sentence as a simple matter-of-fact statement, like 'rain falls from the clouds'.

He tells me of some classic experiments conducted in the 1980s by the American sleep pioneer Allan Rechtschaffen, of the University of Chicago.[3] Rechtschaffen found that if you deprive rats of sleep completely they'll die in about two weeks. If you deprive them of REM sleep they'll die in four weeks. As the rats become more sleep deprived, their behaviour changes. The males become hypersexual. 'They were mounting rocks,' Williams says, dry as sand.

I recall the case of Randy Gardner, who was a teenager in San Diego in 1964. On a whim, he decided to stay awake for as long as he could. That turned out to be an extremely long time: 264 hours, which amounts to just over eleven days. He holds the record for the longest time that anyone has intentionally gone without sleep without using drugs.

By some accounts Randy was fairly unaffected during his non-sleep, although some observers reported that his short-term memory was shot, and that he hallucinated, became moody and paranoid. He was monitored during his record attempt by a US Navy psychiatrist, John Ross.[4] This is from Ross's report of Gardner on day eleven:

> Expressionless appearance, speech slurred and without into-nation; had to be encouraged to talk to get him to respond at all. His attention span was very short and his mental abilities were diminished.
>
> In a serial sevens test, where the respondent starts with the number 100 and proceeds downward by subtracting seven each time, Gardner got back to 65 (only five subtractions) and then stopped. When asked why he had stopped he claimed that he couldn't remember what he was supposed to be doing.

He was, however, fine after he'd caught up on his sleep, and showed no long-term problems.

We all know how lack of sleep makes us moody. Ellen MacArthur's support team saw this regularly during her round-the-world attempt. It happens because the connection starts to break down between the decision-making parts of our brain in the prefrontal cortex and the mysterious and intimidating amygdala, which is the master of our fear and emotions. But the sleep-deprived rats didn't die of moodiness.

'In the totally sleep-deprived rats,' Williams says, 'they die in a very bad state, their temperature is falling, their appetite has gone up but they're losing weight, they lose their fur and the bowel disintegrates.'

I'm at a bit of a loss to respond to this. Kryger helps me out. 'They have a horrible death. And it seems to be a metabolic death. These poor rodents were under horrible stress.'

Once you get over the horror of the experiments, in which rats were housed in cages with rotating floors that would plunge them into water if they fell asleep, it's interesting to note that merely depriving them of REM sleep still caused them to die, if not so quickly. So what does this tell us about the function of sleep?

There are some hypotheses that it is about the practising of complex motor behaviours; we know sleep is essential for consolidating memories[5] and has a function in restoring cell function.[6] This might be why babies have more REM sleep, because they are learning to control their movement.

'Babies spend twelve hours a day in REM sleep,' says Kryger. 'But we don't know if they dream, or if they do, what they're dreaming of.'

Adults too vary in the amount of REM sleep they get.

'We see patients on antidepressants, which dramatically reduce or in some cases abolish REM sleep,' says Leschziner. 'We don't know how important the function of REM sleep is in later life. It may have a very important function in babies or children that is of less significance in later life.'

It may be to do with processing emotions experienced in the day. There remain some fundamental unknowns and we don't know whether the functions of different stages of sleep alter as we go through life. 'It may be that we were designed to die at forty, so the function of REM sleep after forty is of no consequence from an evolutionary perspective,' says Leschziner.

We might not know the function of REM sleep, but we all know about one of its most celebrated consequences.

Read these words: 'Scrambled eggs – oh my darling how I love your legs.' Did you automatically insert a melody? No? How about this: 'Yesterday – all my troubles seemed so far away.'

The melody for the most recorded song of all time came to Paul McCartney in a dream in a hotel room in 1964. On waking he knew he had something, and improvised the first lines that came into his head so he wouldn't forget it. When George Martin first heard the demo it had the working title 'Scrambled Eggs'.

Cut to another hotel room, a year later. Keith Richards wakes from a dream with what would become one of the most famous guitar riffs of all time in his head. He picks up a guitar and plays '(I Can't Get No) Satisfaction' into a cassette machine before he goes back to sleep. Richards later said that you can hear him snoring on the tape.

I like the way that both men, McCartney and Richards, are happy to attribute their most enduring works to dreams, as if shirking responsibility for them. There are dozens more examples. The structure of the periodic table of elements came to Dmitri Mendeleev in a dream, and Otto Loewi dreamed the idea that led to the discovery of neurotransmitters and a Nobel Prize.[7] But while we have probably all been inspired by dreams, you have to be in a special place creatively to achieve a breakthrough on the level of 'Yesterday' as the result of a dream. I am, however, counting the ability to dream as a component of the ability to sleep: superhuman dreamers qualify for inclusion in this chapter. Some people really can dream better than others, and the ability has positive, real-life consequences.

Michael Schredl had kept a dream diary from the age of twenty-two. When he was thirty-four, he tried something different. He started asking himself, five to ten times a day, 'Am I dreaming or am I awake?' He would scan the world around him and check for signs that he was in the real world. If there was something incompatible with reality, he would know he was in a dream.

If this sounds like extreme paranoia, or the start of a Christopher Nolan film, it isn't. Schredl was practising a method known to increase the probability of lucid dreaming — the state when you are in a dream but become aware of it, and are able to take control. You may well have experienced it. About 50 per cent of people have a lucid dream at least once in their lives. Sometimes when it's happened to me it's enabled me to stay calm in the dream when something terrifying is happening, like being eaten by a monster, or stabbed, or falling off a cliff. I won't die, I've told myself, because I know this is a dream. In more pleasurable lucid dreams I've been able to fly or levitate (though sometimes these become 'unlucid' and gravity drags me down).

Around a fifth of the population have lucid dreams once a month or more. 'Not only does the frequency of lucid dreaming vary a lot between different people, but so does the ability to affect dream content,' says Schredl, who grew up to become a professor in the sleep laboratory of the Central Institute of Mental Health at Heidelberg University, in Germany. 'So there are naturally good lucid dreamers.' Let's meet one.

Michelle Carr's first lucid dream happened when she was nineteen and a college student, studying psychology at the University of Rochester, New York.

'I was sleep deprived and often took naps after early morning classes,' she says. 'After one morning nap, I had a false awakening where I sat up in my bed, and then realised that my body was still asleep in bed, and I was in fact dreaming. I floated around my bedroom briefly before waking up.'

That was her first lucid dream. After that she read up on them and practised techniques to induce them. She found that morning naps were good times to go lucid: 'I would sometimes be able to wake up briefly from the nap and then consciously fall asleep into a lucid dream.'

This is the wake-induced lucid-dreaming technique. It's about keeping grasp of the transitional stage between waking and sleeping known as hypnagogia, and bringing some conscious awareness with you into the dream state. 'I believe this technique worked for me because I was likely having more REM sleep and a lighter sleep in these naps than I have during the night,' Carr says.

She now has lucid dreams about once a week, and has done for years. She uses them for fun (flying is always a favourite) and if the need arises she will use them to deal with nightmares, 'Such as confronting one recurring monster who ended up representing a friend with whom I had recently had a fight.' During one dream incident, she remembers scrabbling to escape a monster only to realise it was a nightmare, and one she'd been in before. She took control, turned around in the dream and faced the monster. Carr uses lucid dreams as the rest of us might use spare time. She meditates, practises French (without the social anxiety she would have if doing so when awake) and explores her consciousness, to see what it is capable of creating.

Her lucid dreams were a huge influence on the career that she pursued. After her undergraduate degree she went on to do a Ph.D. at the Dream and Nightmare Laboratory at the University of Montreal, and now works on dreams and emotional memory at the Swansea University Sleep Laboratory in the UK. Her years of practice in lucid dreaming, and the fact that she thinks about it a lot in the day too, means she experiences them regularly. She can maintain the lucid state for ten to fifteen minutes, and is able to control herself during that time.

Martin Dresler of Radboud University in the Netherlands, and colleagues at the Max Planck Institute of Psychiatry in Munich, managed to scan the brain of one person who was able to go into lucid dreaming while in the coffin-like confines of the fMRI machine. Based, admittedly, on this single data point, Dresler found that lucid dreaming occurs during REM sleep but shows activity in brain regions normally shut down in that stage of sleep. This, the researchers suggest, could explain why lucid dreamers are able to access cognitive abilities, such as self-control and access to memories, that are beyond us in normal dreams. The biggest difference between lucid and non-lucid REM sleep was seen in the precuneus, part of the brain involved in self-referential processing, agency and first-person perspective.[8]

This fits with Carr's experience of lucid dreaming. In such dreams, she has self-awareness and control over her own movements, but not absolutely over the dreamscape. 'I still run into resistance from the dream,' she says. 'For instance, I am not really able to change the environment around me at will, although I can decide to move through it.'

I've occasionally met dead relatives in my dreams. In these dreams I'm aware that the person I'm with is dead in the real world, not least because they appear younger than they were when they died, but I don't guide the conversation; the encounter takes its own course. This is a common feature of lucid dreams. They aren't always about creating elaborate plots; they can just unfold, with the dreamer as the passive but lucid observer. With dead people – not that I see dead people all the time, but when I do – I don't steer dream conversations round to particular subjects ('Tell me where you hid the diamonds, Grandma!'); the conversations and the encounters are unguided, they would be mundane if it weren't for the fact that one of the characters is dead. When I wake I am left with an odd feeling of pleasure and affection at having spent some time with a loved one who is now dead, even though the experience was entirely imaginary. It is this emotional quality which makes the dream memorable, as we saw in Chapter 2. Lucid dreams are pleasurable, and even beneficial. Evelyn Doll, for example, from the Medical University of Vienna, found that regular lucid dreamers had better mental health than normal dreamers.[9]

It's instructive that the characters who pop up in a lucid dream usually retain their autonomy, because it tells us which regions in the brain are important for consciousness. In Carr's dreams, too, characters remain beyond her control. 'I might approach someone and ask them for information, but they either ignore me or respond in gibberish. Sometimes I am unable to open doors or to approach certain characters,' she says. 'So I think that my techniques still require a lot of practice.'

Michael Schredl's work points to what can be achieved with practice. For researchers there's an extraordinary benefit to studying lucid dreaming: you are able to communicate with people in the unconscious world. The dreamer, of course, is asleep, usually in REM sleep, but as they achieve lucidity, they can signal to the scientist with their eyes. Kristoffer Appel, at Osnabrück University in Germany, has exploited this by teaching Morse code to lucid dreamers, and having them send messages back from the dream world. An eye motion to the left is a dash; to the right is a dot. Appel sends the dreamer arithmetic to perform by playing sequences of beeps that the dreamer incorporates into the dream.[10] The dreamer is then able to compute the answer and signal it back using Morse code.

Schredl has taken this further, and uses the lucid dream as an arena for training. This leads to the remarkable proposition that you could intentionally practise something while sleeping, in order to improve your skill. In one study of 840 athletes, 57 per cent said they'd experienced at least one lucid dream in their lives, and 24 per cent had one at least once a month.[11] Nine per cent of the lucid-dreaming athletes practised their sport in the dream, and said their athletic skill improved as a result. It's a small sample, and entirely anecdotal, so Schredl and colleagues decided to try and test the idea. They brought lucid dreamers into the sleep lab and had them practise darts while dreaming. Yes, darts. We are in a Martin Amis novel written by a neuroscientist.

Schredl and his team had a group of test subjects play darts, then fall asleep in the lab. Volunteers who were able to go into lucid dreaming performed three consecutive left–right eye movements to signal the onset of dream control. First they had to 'assemble' — that is, dream into existence — the board and the darts necessary for the trial. The boards used were the kind with a red bull's eye at the centre, surrounded by nine concentric black and white rings. Then they began to play darts. The trials required subjects to throw six sets of five darts, and to signal to the waking world after every fifth dart.

At the end of the thirty throws, the dreamers had to try and wake up. They then provided the researchers with a detailed dream report. The next morning, lucid dreamers and control subjects who had not tried to practise darts in their sleep were tested again with a real dartboard, and scored for their ability.

Just as with any kind of practice, distractions detract from its quality and efficiency. Some lucid dreamers were able to practise darts without distraction, but others had to deal with obstacles, such as a pesky dream character interfering with the task ('The doll kept throwing darts at me') or changes to objects in the dream environment ('At some point I threw pencils'). Sometimes the dream got away from them, and the dreamer's grip on lucidity started to fail ('I noticed it was getting somewhat unstable . . . I performed another eye signal . . . I managed three or four more throws and then I woke up'). The scientists tried to account for distractions in the dream reports obtained from the dreamers.

You can imagine the difficulties in obtaining a good sample size under these trying conditions, and Schredl emphasises that this is a pilot study. Given that caveat, they found that if dreamers weren't distracted, their darts score improved after they practised in the dream. Given that almost a quarter of German athletes apparently have regular lucid dreams, Schredl wonders if dream practice might become part of athletic training.

He told me about reports from amateur athletes who improved their skills during lucid dreams. There is a springboard diver who used lucid dreams to practise twists and somersaults. She could slow down time in the dream so the sequence unfurled more slowly, allowing her to understand the movements going on at each point. There's also a snowboarder who practises tricks that he can't yet do in the real world. The practice has helped him improve, he says.

It makes me think of the training philosophy of marginal gains. It's arguably been responsible for some of the huge success in British cycling over the last decade or so. The idea is that you try to improve every possible factor so as to reap a greater benefit. Such as painting the floor of the team truck white, in order to better spot any dust that might impair bike performance, or finding the best pillow for sleep and bringing it to team hotels.

On its own, each adjustment permits a tiny, maybe unnoticeable improvement. But make enough of them and the gains mount up. Sports psychologists already place a lot of emphasis on getting a good night's sleep. Should dream-training be added to the list?

For sports such as cycling, where strength training is key, it probably won't work, says Schredl. 'But for more artistic sports, like platform diving and freestyle skiing, this does seem plausible.' It's why mental training and rehearsal while you're awake has a benefit: think of the track and race-line memorising we saw in Formula 1 drivers. Direct transfer of information during sleep has been achieved in rats, but this has required the highly invasive implantation of electrodes. Dream training may work only if you can achieve a lucid state, but it's possible to learn this, and at an elite level, any small improvement is worth pursuing.

LeBron James is one of the greatest basketball players of all time. He is a double Olympic gold medallist, has won the US National Basketball Association championship three times and has been voted Most Valuable Player four times. There are many reasons for his phenomenal athleticism and his success, but one jumps out at me: he sleeps eleven to twelve hours per night.[12]

Cheri Mah, who now works in the Human Performance Center at the University of California, San Francisco (the lab where ultrarunner Dean Karnazes was tested), investigated the effect on college basketball players of longer sleep. She and her colleagues recruited eleven men from the Stanford University team, with an average age of nineteen, and trained them to extend the amount of time they slept. The men all managed to sleep longer (by an average of almost two hours), and their sprint speed, shooting accuracy and free throw percentages all improved. Extending sleep time, Mah's team concluded, improves athletic performance, reaction time and sprint time, as well as mood and vigour. Peak performance can occur only when sleep is optimal, she suggests.[13]

There's only one LeBron James, but anyone can extend their sleep time. You just need to make sure you don't drink alcohol or caffeine too late in the day (and preferably not at all), nor should you eat dinner too late. Both things keep your metabolic rate high and can disturb your sleep. Make sure your bedroom is cool, dark and quiet, and that you won't be interrupted or disturbed. Don't go to bed too late, and try not to binge-read on electronic devices, or check work emails before you do. Wind down at least half an hour before bed, crank the lights down low, read a novel. Embrace sleep as your restorative friend rather than something to resist.

There will, however, always be the resistors. Some will sleep six hours or fewer from pressure of work, and some will insist they don't need sleep. It's at this point in the chapter that we bow to the inevitable and mention Margaret Thatcher and Donald Trump. Both politicians are often cited for their ability to get by on little sleep. 'Trump brags that he only needs four hours a night,' Meir Kryger told me, 'and it shows, in my opinion, because he has many of the symptoms of sleep deprivation.'

These symptoms include moodiness, lack of alertness, confusion and problems making decisions. In Trump's case, the garbled 'covfefe'[14] tweet springs to mind. Long-term lack of sleep has been linked to an increase in risk of stroke and diabetes, and can cause depression and weight gain. The US National Sleep Foundation's latest, updated sleep duration figures recommend seven to eight hours' sleep per night for adults.[15] Recent work at the Sleep Research & Treatment Center, Penn State College of Medicine in Hershey, Pennsylvania, suggests there are side-effects to short sleep.[16] If you are already at risk from cardiovascular disease, short sleep can increase the risk. Julio Fernandez-Mendoza also found that there are some short sleepers who might not even be aware they are short sleepers. Typically they think they get more sleep than they actually do. 'In other words, they are not aware that when they spent eight hours in bed and think they got seven and a half hours they actually got six,' he says. These kinds of short sleepers don't seem to suffer any ill-effects, such as hypertension, diabetes or depression, but they do experience a decrease in processing speed. This is part of cognitive function that determines how quickly you are able to understand, react to and complete a mental task.

Some of the claims made by people who say they only need four hours' sleep are bravado; they are what business people and politicians (people in these roles are especially prone to the tendency) think they ought to say. 'Some people are short sleepers and are macho about it,' says Kryger. 'They don't perform really well, they think sleep is a waste of time. They'd rather be awake or making money and being productive rather than sleep.' On the other hand there really are people who don't suffer the cognitive hangover most of us get if we miss out on sleep. Businesswoman Martha Stewart seems to be one of these.[17] She has complained that there's not enough time in the day and says she needs only four hours a night.

'Some of the claims people make are true,' says Kryger. Some short sleepers seem to escape the negative consequences. These are the people we'll investigate now.

The Sleep-Wake Center at the University of Utah, like

most sleep units around the world, mostly sees people who have problems sleeping. They are busy fielding patients with sleep apnoea, daytime drowsiness, restless leg syndrome and, of course, insomnia. But in the early 2000s, its director, Christopher Jones, started pondering why some people are morning larks, and some skew much more to the night-owl end of the sleep spectrum. He realised that understanding this better might inform treatments for people who don't sleep well, so he put out a call for volunteers for a study. He was interested in habitual early risers, and a 68-year-old woman got in touch. The woman told Jones that she needed only six hours' sleep, that she had been like that as long as she could remember, and that it had never her done any harm. What's more, her daughter was the same.

Jones was intrigued and fascinated, especially that the daughter was the same – it hinted that there might be something genetic influencing their sleep. In the early 1990s Urs Albrecht and colleagues at the University of Fribourg in Switzerland had found *Per2*, a gene in mice that regulates the circadian clock. There's a version of the gene in families that have advanced sleep-phase syndrome. If you have this, you sleep eight hours but you are skewed wildly out of kilter. These people are extreme morning people, going to bed at six or seven in the evening and waking at three or four in the morning. The *Per2* gene, on chromosome 2, influences the circadian pacemaker the body sets its clock by, and variations of it have been linked to certain cancers.[18] Ever since this variant had been discovered, says Jones, he'd been hoping to come across a family with highly skewed or unusual sleep patterns. With this mother and daughter pair, he seemed to have found it.

'I had never heard of "natural" short sleepers before,' he says, 'and at first just thought they were just very morning people.'

However, he had the pair keep sleep logs and wear wrist monitors called actigraphs which recorded the amount of movement and activity they made. The results confirmed that the pair were going to sleep late (around 10 p.m.) as well as getting up early (at 4 a.m.), which is not generally considered part of the usual 'extreme morning person' routine.

Jones took DNA samples from the pair and sent them to Ying-Hui Fu, a colleague at the University of California, San Francisco. Fu specialises in the biology of myelin, which is the fatty insulating material around nerve cells; and sleep behaviour. She has a particular interest in genetics, and in the extreme larks of the sleeping world.

When she looked at the gene sequences of the mother and daughter pair referred from Utah, Fu found a mutation in the DEC2 gene on chromosome 12. Suspecting that it was this mutation that was causing the women to wake up so early, Fu made transgenic mice and flies carrying the mutation. The resulting mice slept about an hour less than normal mice, and the flies two hours less. Fu, who published her findings in the journal Science in 2009,[19] says if DEC2 was available in a pill, she'd take it so she'd have more time in the day.

Louis Ptacek is a neurogeneticist who works with Fu at UCSF. The key question about people with the DEC2 short sleeping mutation is whether they gain all the restorative benefits of sleep in only six hours.

'Unfortunately, we know so little about sleep that this question cannot be answered at this time,' was his response. The issue he says, is whether these people 'need' less sleep. In other words, do they accumulate while awake less of whatever it is that must be cleared during sleep? 'Do they have the same "burden" from being awake for a certain number of hours but sleep more "efficiently"?'

We don't yet know.

Since finding the mother–daughter pair, the team have gathered genetic information from many more families of short sleepers, and according to Ptacek they have discovered at least one or two, and maybe three new human sleep genes or mutations. The team is focusing now on people with this short sleeping ability, and will be able, he says, to get at some of these questions: 'All elements of sleep are genetic. And they are all subject to environmental influences.'

So there are genuine short sleepers. People who naturally get by on less, not those who push themselves because of work or because they have to care for a relative. What are the personality traits of these people?

To find out, Timothy Monk and colleagues at the University of Pittsburgh Medical Center in Pennsylvania recruited short sleepers for a study, making sure they admitted only natural short sleepers. Those selected (nine men and three women) slept for an average of 5.3 hours per night compared with 7.1 hours for a control group of people matched for age and gender who were not short sleepers.

All the subjects completed attitude-to-life questionnaires. If you know someone who is a short sleeper, or if you are one yourself, consider for a moment the personality traits you associate with them. Monk's results showed that in their small sample, the short sleepers were more energetic and aggressive than the normal sleepers. They were also less anxious and more ambitious. All these traits seem to fit the stereotype of the short sleeper we hear about. In contrast to earlier studies, Monk's team found no evidence that short sleepers were necessarily more extrovert, but concluded that there is some evidence for 'subclinical hypomania' in short sleepers. This is basically an elevated mood that when it becomes problematic can tip into bipolar disorder.[20]

Remember that the short sleepers in this study were not being driven by work. The team had to exclude many applicants who put themselves forward as natural short sleepers but who were found to be sleeping 'unwisely' – curtailing sleep for work or caregiving purposes, or because they were unwell either physically or mentally.

'Every now and again I'll encounter someone who doesn't sleep much in terms of what we consider to be normal, but is quite productive,' says Meir Kryger. 'What we'll never know is how productive they could've been had they actually slept the right amount of time. That's the dilemma.'

Productivity is one thing, but the evidence is clear that it's not good to force yourself to sleep less than about seven hours. Napoleon Bonaparte's pronouncement about the hours of sleep people require – 'six for a man, seven for a woman, eight for a fool' – is as backward as it is sexist. Even for those people who genuinely get by on less sleep, such as those being profiled in Ying-Hui Fu's lab, the long-term costs will eventually outweigh the short-term benefits. We can say this because sleep deprivation seems to make people more vulnerable to developing dementia.[21] Several lines of evidence suggest a reason. For one thing, the clean-up cells deployed by the brain are overactive when you are sleep-deprived. Michele Bellesi of the Marche Polytechnic University in Italy interrupted the sleep of mice for five days in a row, and found that cells called astrocytes that perform a pruning function in the brain were more active, as were microglial cells, which function to locate damaged cells. Other experiments in mice have shown that sleep helps the brain purge itself of debris[22] and that if you prevent animals from getting deep sleep, then amyloid proteins – the hallmarks of Alzheimer's – build up in the brain.[23]

Remember that Margaret Thatcher suffered from Alzheimer's late in life. It seems we are now able to say that her habit of sleeping only four hours a night caught up with her – that there was a causal link between her sleeping and her dementia. 'Animal tests would suggest it's causal,' says Adrian Williams. 'Mice get Alzheimer's if they don't sleep.'

While it's clear that a full understanding of all the benefits of sleep, let alone the mechanisms of how it functions, will take many years, we know that a good night's sleep is essential for our health and well-being. Get some rest. Your happiness depends on it.

11

HAPPINESS

Happy is what you realize you are a fraction of a second before it's too late.

Ali Smith, *Hotel World*

We all look for happiness, but without knowing where to find it: like drunkards who look for their house, knowing dimly that they have one.

Voltaire, *Notebooks*

Until her life changed radically and for ever at the age of forty-one, Shirley Parsons was a successful solicitor in Exeter, south-west England. Her husband ran a nearby farm. Beef and sheep. They are still married, though these days she doesn't see him all that often.

I chatted with Shirley on email for a number of months, and during our conversations I was struck by her thoughtfulness, and by her extraordinary reserves of strength and resilience, though when I put that to her she said, 'Most people would use the words stubborn and awkward!' What also struck me was her approach to life. 'I've come to the conclusion that my brain's default setting is happiness,' she said on one occasion.

This opened the door to a discussion on the nature of happiness. I asked her if she could define it. There was a long time before a reply came. 'To my mind there are different types of happiness,' she said eventually. 'A general happiness with life, happiness due to anticipation of a specific event such as a wedding or party, and happiness created by a particular moment such as passing an exam.' Her conclusion: 'I can't give a single definition for happiness.'

Well, she's not alone. Thomas Jefferson didn't define it in the US Declaration of Independence, instead only guaranteeing the right to pursue it. Happiness is at least as slippery a concept as intelligence, and the supposed routes to achieving it are labyrinthine, as we'll see.

At this point Shirley and I had only ever been in touch by email and we both agreed it would be good to put a face to a name. So we arrange a visit. Now fifty-five, she still lives in Devon, and I drive down to see her. Her house is in a village on the edge of Dartmoor, a large, granite-studded national park which I usually think of as rugged, if not bleak, but today it is incredibly green. It is beautiful, and skimmed by swallows and swifts. As I'm driving, a crow chases a buzzard right in front of my car, and I see the raptor turn its head in mid-air, irritated, and yell at the aggressor. The countryside is roasting. It's the hottest day of the year so far, and I arrive at Shirley's place in the middle of the day. There is a freshly mowed lawn leading up to the house, and rows of plant pots with flowers. The doors are open; I have phoned ahead and am expected, so I go in through the front room. I catch sight of Shirley's two university graduation certificates on the wall; on the mantelpiece is her wedding photo. I go into the bedroom, which is dark, curtains drawn. The bed is elevated, with scissor legs; it is a medical bed. There's a pouch for an intravenous drip hanging on a stand.

Shirley sits watching the tennis. She doesn't turn around to say hello, because she can't move. Nor does she speak, because she can't speak. She has been paralysed from the neck down for 14 years and 5 months. She has locked-in syndrome: her mind is intact, and indeed thriving, but her body has long since closed down.

It started on a Sunday morning in 2003. Shirley woke up with a really bad headache and a sensation of vertigo. She decided to stay in bed, but dragged herself up in the afternoon to help with feeding on the farm. 'I suddenly felt dizzy so I sat down on the hay,' she recalls, 'and the next thing I knew was when I woke up in intensive care two weeks later.'

Face to face, she can answer yes–no questions by looking up (yes) or from side to side (no). If I ask a more complicated question, she answers using her computer, which she operates using a cheek switch, and specialist software called EZ Keys. This is how she's been emailing me.

Like Carmen Tarleton, whom we met in the chapter on resilience, Shirley was in a medically induced coma for those first two weeks. Doctors discovered that she had a mutation in the gene for factor V, the stuff that helps blood clot. The single-letter misprint in the code meant that one of her factor V proteins has a glutamine amino acid instead of an arginine. This gives her thrombophilia, the tendency for her blood to clot more than it ought to. The type she has is called Factor V Leiden, after the Dutch city where it was discovered. About 5 per cent of people of European descent carry this mutation, which increases the risk of developing deep-vein thrombosis, and which in rare incidents can cause a detached blood clot to travel to the brain. When this happened to Shirley it caused brain stem bleeding, and a massive haemorrhagic stroke. She was in a rehabilitation centre in Exeter for more than a year after the illness burst. She was told that she wasn't expected to live.

In her room, I move into her line of sight. Her eyes move on to me. She is in a wheelchair, wearing lots of layers, and I wonder that she's not sweltering in this weather. Her hands are on her lap but covered in closed cotton sleeves. Her face is flushed and sheened. Her mouth hangs open, which is the default setting now she's paralysed. It gives her a hospitalised look. It's the look you see on people in intensive care, on people in comas. On people in vegetative states. I remember it on the face of my grandmother after she'd fallen into a coma. You see people who look like this and you assume that they are paralysed in their minds, too. Labels such as 'vegetative state' don't help. Clearly I know that's not the case with Shirley, but it's still disconcerting, to see her face to face, unable to move on the outside but knowing that she's buzzing, on the inside. 'I think too much,' she's said. She's also said she is happy with her life. I then asked her, apologising if it was a strange or even offensive question, whether there was a sense that she was happier now than before she became locked-in. She found the question easy to answer. Easier, she said, than when I asked her to define happiness.

'Although on the face of it, it seems a ridiculous question, when I thought about it I realised that it wasn't, because rather bizarrely I think that I am happier,' her emailed reply said. 'Before the stroke my life was noisy and hectic but now most of the time it's quiet, peaceful and calm. Over the years I've grown accustomed and become content with my life.'

You often hear people saying they want to make their lives simpler. We complain that there are too many distractions in the world, in our busy lives, and we think we'd be happier if we got rid of some of them. Maybe it's true, maybe we would be happier. Later we'll meet a man who has got rid of all his possessions apart from only seventy items. He has fewer material goods than anyone I know, and seems happy. We'll meet a professor at one of the world's top universities who gives away everything he earns above £25,000 (about $32,000): he says he's far happier than before he took the donation pledge. Perhaps getting rid of stuff really does make us happy. According to a popular search engine, the happiest person in the world is Matthieu Ricard, a French academic-turned-Buddhist monk who is an interpreter for the Dalai Lama. Buddhists don't have much in the way of material possessions and they tend to be, if not radiantly happy, then at least suffused with well-being.

I've come to visit Shirley because she's got rid of almost everything. She didn't choose to, granted, but she lives a life stripped down to the barest minimum: just a brain, looking out. So I want to gain an idea of what her life is like. Perhaps it's not so unbelievable that she could be happy. But happier, even, than before?

'The great snare of the psychologist,' said William James in 1918, 'is the confusion of his own standpoint with that of the mental fact about which he is making his report.' What he called the psychologist's fallacy is, in other words, the assumption that we know what someone else is experiencing. James was the father of psychology but it seems to be a tendency that we could still usefully bear in mind. Consider the case of someone with locked-in syndrome. To their family and friends, they have lost everything. They are completely dependent on others for all their needs. Many of us simply assume life wouldn't be worth living. Martin Pistorius, a South African author and web designer, became locked-in as a teenager, and was paralysed for more than a decade. He recovered, and later described hearing his mother tell him, thinking that unresponsive meant unconscious, 'I hope you die.'[1] He said his parents had been told that he was a 'vegetable' and that they should wait for him to die.

If you were fully conscious but paralysed, can you even imagine how you would feel, what you would miss? I'd asked Shirley this. One thing, she said, was toast. She missed crunching on a piece of toast. I ask her the same question again when I visit her, and she starts laughing. She laughs a lot. Although paralysed, she isn't completely silent. Unlike many people with locked-in syndrome, she has regained some head movement, and can swallow and make noises. It means she can be fed with a spoon rather than a tube, and it means she can laugh. Her laughter is a funny kind, that's for sure, but I hear it a lot as I sit with her. This particular laugh is hard to evaluate at first: I worry she is crying, or even choking. But I realise it's definitely amusement. The panting laugh goes on as she spells out her reply on the computer. What does she miss? 'Being able to talk.' Most of all she misses chatting. 'The carers don't talk, and with good reason,' the message goes on. 'I don't suffer fools.' Because she would tell them off? She laughs some more, eyes raised to the ceiling: an emphatic yes. The carers do talk to her, of course. But perhaps what Shirley means is she misses banter and discussion; conversation, argument. She was a solicitor, after all.

I first began thinking about people with locked-in syndrome after meeting Marie-Christine Nizzi in the department of psychology at Harvard University. Nizzi's office is at the top of William James Hall, a famous and appropriately named highrise in Cambridge, which has a fantastic view over the city. Sat up there, she told me the parable of the Chinese farmer and the horse.

There was once a farmer who was offered a lot of money for his beautiful white horse, but he refuses. It's the only thing he has, he says, he doesn't want to part with it. The offer is rescinded and the villagers in the neighbourhood tell him he's crazy for missing out. A while later, the horse escapes, and now the farmer has lost even the horse. The villagers jeer at him but he shrugs. Who knows what will happen, he says: it could be good, it could be bad (another name for this parable is the equanimous farmer).

A few days later the horse returns, leading a herd of wild horses. The farmer is now rich, and the villagers urge him to celebrate. But he doesn't, trotting out his usual line, 'It could be good, it could be bad.' A while later his son breaks his leg trying to tame the wild horses, and the villagers (they are a miserable lot) cluck at him that he wishes now he hadn't been so lucky with the horses. It goes on like this. War breaks out, and all young men are conscripted, apart from the farmer's son, who is exempted with his broken leg.

The moral of the story, said Nizzi, is that we shouldn't jump to judge the good or bad of an event. 'There can be silver linings. A lot of people project how they would feel in those circumstances, be it the lost horse or the loss of money.' She felt this was something that was being missed widely, in psychological evaluations, and started working with people with locked-in syndrome. But how do you know if they're happy or not? 'How about we ask them,' she said. The way she said it, so simply, irrefutably, made me realise what an incredibly obvious question it was, but Nizzi reassured me that many people with locked-in syndrome are never asked about the quality of their inner life. We just assume it's horrendous so we don't ask. Also: we don't want to know. So that's what she did. She asked them about their perspective.

Nizzi calls it going from the armchair to the wheelchair,[2] which is also what the Afghan war veteran Dave Henson in Chapter 5 did to gain perspective from the other side. When Nizzi surveyed patients with locked-in syndrome, 'What I found was that they have a much more positive quality of life than what we project from the outside.' Their resilience, she says, allows for a satisfactory quality of life. For example, they want to be resuscitated in case of an accident, when nurses, doctors and even family members sometimes presume that the patients wouldn't want to live like this. 'The patients report being happy, or being satisfied with their lives. The majority are happy.'

Nizzi's work follows that of Steven Laureys, who runs the Coma Science Group at the University of Liège in Belgium. In 2008 Laureys decided to conduct a quality-of-life survey of people with locked-in syndrome. Sixty-five patients were included in the survey; forty-seven of them professed themselves happy, and the rest responded that they were unhappy.

Things that made them unhappy included anxiety and their lack of mobility, as well as Shirley's biggest regret: the loss of speech. The longer they'd spent in the locked-in state, the happier they were, but still, 58 per cent of the people responding said they did not wish for resuscitation in the event of a heart attack.[3] Laureys concluded that perhaps the happy locked-in people had succeeded in recalibrating their lives.

I accept that some locked-in patients are content, even happy. But surely no one locked-in feels better about their lives now? 'I asked that exact question,' says Nizzi. 'The answer was yes. They knew themselves better, and they found their life's meaning more strongly now than they had before.' Being locked-in forces you to halt. It forces you to look for a meaning that you may not have sought before. It's only because I've heard this that

I dare ask Shirley if she's happier now than before her stroke.

Tim Harrower is Shirley's consultant neurologist at the Royal Devon and Exeter Hospital, and saw her when she was in rehab there after her stroke.

'The big breakthrough with locked-in patients happens when you work out a way to communicate,' he says. 'It changes everything, because they can communicate their needs.' You might have an itch, or you're in pain, or you just want the damn TV turned off. Then it's a case of adjusting to what you can and cannot do, and finding something cerebral to do. 'Being able to accept your situation and adjust to your limitations is the biggest problem really,' Harrower says. 'That doesn't happen straight away, it can be years down the line.'

Shirley had been locked-in for about five years when she decided to begin a degree course. She was missing her job; she was bored, basically. 'It was the loss of my job which prompted me to study,' she says. 'I was too young to retire and needed something intellectual to fill the void.' She applied to the Open University. First she took a BA in social sciences with politics. She remembers writing an essay about Tony Blair's style of government. She graduated in 2010. Then she took a postgraduate certificate in the same subject, graduating in 2012. It must have been challenging, I offer. 'The main challenge,' she says, dryly, 'was that it took me ages to type assignments.'

Shirley has always been a dogged woman. She tells me about her response to the disastrous outbreak of foot-and-mouth disease which hit the UK in 2001. This was before she was locked-in, when she was working as a solicitor. Their animals had not contracted the disease, but those on a neighbouring farm had, and it was government policy to cull animals on properties adjoining those infected. Deeply disturbing images were shown on TV of pyres of burning cattle. 'After initially being upset I went into fight mode,' Shirley says. 'To cut a long story short, I was successful in preventing all the cattle and the vast majority of the sheep from being killed.'

Shirley wears glasses, and the left-hand lens of the glass is a bit steamed up. There is a humidifier pumping out steam on her left side, next to where I'm sitting, and maybe that clouds up that side of her glasses. So really I just have her right eye to look at, and I find it fixes me with an intensity I can't hold all the time. I have to look away, at her DVDs (Manchester City Champions; Bridget Jones; *Orphan Black*; *12 Years a Slave . . . 12 Years a Slave*!) and at a large collage of photos from work colleagues with the legend 'C'mon Shirley! You can do it!!' I sit there, babbling away, aware that soon I'll drive back to my life, knowing that the majority of the rest of her life will be spent in this bungalow, in this dark little room, and she knows I know that, and here we are talking about how happy she is.

I'd heard about William MacAskill by reputation. Not only was he one of the youngest professors in the world when he was appointed to the University of Oxford's department of philosophy, but he made the decision in 2009 to give away most of his income over the course of his life. Everything he earns over £25,000 he gives away. I was amazed and impressed by this, and not simply by the generosity. How does he manage to live in Oxford, I wondered. It's not cheap, it's almost as pricey as London. Before I met him I half-wondered if he would be like an ascetic monk, barefoot and in sackcloth.

MacAskill is also known as one of the founders of the effective altruism movement, which uses science and evidence to discover the most effective ways to donate money. He started Giving What We Can, an organisation that encourages people to pledge at least 10 per cent of their income to charity, for life, and to give it where it will do the most good. Giving What We Can launched with twenty-three people, and now has nearly 3000 members who have together pledged $1.4 billion. I meet him at the Centre for Effective Altruism in Oxford, where he is director.

To my slight disappointment, he is not monk-like. He is tall, neat, dressed in a clean T-shirt. He even has a smartphone.

His decision to give away so much of his earnings came after meeting Toby Ord, another philosopher here who has made a similar pledge. The decision was made because of an understanding of how money contributes to happiness, and the difference in happiness he can make by donating.

There's a famous paper by economist Richard Easterlin that is often cited in discussions on happiness.[4] In 1974, Easterlin, now at the University of Southern California, published data purporting to show that as a country's GDP grew – for example in Japan, spectacularly, after 1945, or in the US, during the course of the twentieth century – the happiness reported by citizens in those countries did not increase. This is known as the Easterlin Paradox, because it must be a paradox that more money doesn't make us happier, right? I'd heard of this concept, and thought it was a true phenomenon. I thought it was evidence that we were trapped in a system of economics that was driving ever more growth without increasing well-being. MacAskill disabuses me of this.

In 2008, Betsey Stevenson and Justin Wolfers at the University of Pennsylvania looked again at the data.[5] They found there was no paradox. As income increases, so does happiness. It's very clear, and the relationship holds for a wide range of countries: the US, China, India, Japan, Germany. What does this mean – that money does buy you happiness after all? Well, sort of. The Stevenson and Wolfers data shows that as you double your income, you double the amount of happiness you report. The relationship stays the same at all points along the graph. That means if I'm in rural China and am earning $1000 per year my happiness is boosted by the same amount if I get a pay rise of $1000, as it would be if I was in the US earning $80,000 and get a rise to $160,000. If I'm earning eighty grand and get a twenty grand rise – itself pretty unlikely, right? – my happiness doesn't really improve by much.

This kind of measure of happiness is called lifetime evaluation. Basically you ask people, 'On the whole how are things going for you these days?' On this measure, more money does make you happier, but because of the doubling effect, this only works up to a point. 'Above a certain level of income, earning even more makes a negligible impact on your well-being,' says MacAskill.

There is another main way people measure happiness, and that's called experiential sampling. In this method, you buzz people randomly on their mobiles throughout the day, and ask them, 'How happy are you right now, on a scale of 1 to 10?' In this way researchers are able to assess what are the most enjoyable activities. Can you guess what people get most happiness from? 'Sex wins by a huge margin,' says MacAskill. 'People must've been interrupting it to give a 9 out of 10.'

Experiential sampling is interesting because it gets around what's known as the focusing illusion. Another philosopher, Michael Plant, told me about this. 'You and I imagine how much fun it would be to be Kanye West driving around in his Mascrati, but he's just annoyed about being stuck in traffic. So there's a gap between how we imagine other people's lives, or our future lives, and what they're like when we live them.'

In other words we look at Kanye's car and think that's cool,

I'd like that, but we don't focus on what the person in the car is experiencing. We're misled by our desires. Incidentally, talking about moments of transient happiness, there's a sign on the wall of the pub where we meet saying, 'This is where Bill Clinton did not inhale when he was a Rhodes Scholar at Oxford'.

Back to MacAskill. 'On the experiential sampling measure you get no increase in happiness beyond $75,000 per household,' he says. He's referring to a paper by Daniel Kahneman and Angus Deaton of Princeton University that analysed 450,000 such experiential responses from 1000 US citizens.[6]

So let's recap. If you measure happiness as a whole, using the lifetime evaluation method, you find that happiness increases steadily with income. If you use experiential sampling, the measure of moment-to-moment emotional happiness, you find it never increases after a certain, surprisingly low, level of income.

'In the UK, that's about £25,000 for a single person,' MacAskill says. This seems to be a good time to bring up the fact that he lives on £25,000 in an expensive city. 'I give away everything over £25,000, and that still puts me into the richest 50 per cent in the country. And if the 50 per cent who are poorer than me can still live, then surely I can as well.'

There's not much I can say to that without sounding like Scrooge, or more correctly, like the London-based, middle-class journalist I am. But I plough on anyway. How does he afford somewhere to live?

'I pay a little over £500 a month shared rent, I live in a house with ten other people. It's great, I'm much happier in a shared house than I would be living alone or just me and my partner — because of the sense of community.'

MacAskill is very good natured about letting me know how woefully middle class I am. I guess he's had a lot of practice at it. 'Ninety-seven per cent of people in the world live on less than that amount of money. So it's kind of a strange thing when people are like, "How could you live on that?" It's a very middle-class thing because 97 per cent of the world do it. They get along fine.'

On the shelf there's a vintage black-and-white photo of a stern-looking young man. To divert attention from how selfish and consumer-driven I am, I ask who the man is.

He is Vasili Arkhipov, credited with saving the entire world. Arkhipov was a Soviet naval officer who on 27 October 1962 was on board a submarine awaiting orders from Moscow. It was the very pinnacle of the Cuban missile crisis. The submarine's captain, having not heard from Moscow, made the decision to launch a nuclear strike. Authorisation required the three senior officers on board to agree unanimously, and one other did, but Arkhipov refused, and argued the captain down from the launch. Had it gone ahead, the US would have almost certainly responded in kind, and all-out nuclear war would have broken out. An aide to John F. Kennedy later said it was the most dangerous moment in human history.[7] Arkhipov's picture is up here in the Centre for Effective Altruism as inspiration.

MacAskill says his life is much improved since he took the pledge: 'I'm much happier than I was before. I think I have a pretty good idea of what my life would've been like otherwise.'

The happiness seems to come from doing something he derives huge satisfaction from, something that also has huge moral payback, in that giving money away in the way he does improves the happiness of others far more than it depletes his own. 'I don't have a car, I don't have many material possessions, I think they would make my life worse. If I had a yacht I'd be stressed about having this thing I've not got to worry about.'

Quite frankly I'd love a yacht, but I'm not greedy and would be happy with a rowing boat. I'm briefly adrift, considering the question of which I'd prefer, when MacAskill says, 'One of our employees only has a hundred possessions.'

I ask him to repeat this.

'He only has a hundred possessions. He's a minimalist.'

My first response is not to express consternation or admiration, but to ask how many pairs of undergarments this man owns.

'He has two pairs of pants,' says MacAskill.

I have to talk to this guy.

Pablo Stafforini is a chisel-jawed, tousle-haired Argentinian tango-dancer and philosopher living in Oxford. He works as a research analyst here at the Centre for Effective Altruism. 'I don't aim at a particular number,' he says. 'Instead, I try to follow the heuristic of getting rid of as much stuff as possible.' His possessions currently number around seventy. He thinks people are 'biased' towards resource accumulation – that effectively we are trapped in a consumerist system and an evolutionary state that tilts us towards wanting and accumulating too many things: 'This heuristic is an attempt to correct for that bias. The goal is just to live a richer, more fulfilling life.'

As a philosopher he was familiar with the findings from psychological studies we've just been hearing about: that spending money on experiences often makes people happier but spending on material goods usually doesn't. He was also finding that his own possessions – including a piano and a library of some 3000 books – were becoming a nuisance. It had been a pain, and an expensive one, to ship his library from grad school in Toronto to Oxford. So he committed to getting rid of almost everything, and he announced as much on Facebook. He sold the books and the piano. Now if he wants to buy something, it is subjected to a strict test: 'Do I really need this? Will it make me happier?'

He owns some clothes, a pair of boots, shoes and trousers for tango, a decent laptop and a few bits and bobs. I've never met any adult who has so little. He also donates a substantial percentage of his income to charity. Some people think he's crazy. 'These people mostly believe I suffer from a cute or funny type of madness,' he says. 'Overall, the reaction I get is very positive.'

He allows that he *thinks* he is happier now he's shed his worldly goods. He is cautious about concluding that he is happier now he is a minimalist, but that's heuristic-following philosophers for you, even fun-loving, tango-dancing ones. 'It is worth remembering that the psychological literature teaches us that humans are very unreliable at recollecting and aggregating past experience,' he says. In other words, he *feels* happier but won't commit to asserting that to be true.

When Shirley Parsons said her default setting was happiness, what did she mean by that? She seems to be referring to the common-sense view that some people are naturally happier than others. We all know people who are always positive and sunny, and who return to a stable level of happiness after a knock. This has a name in psychology. It is hedonic adaptation, and crucially it works in both directions: it means you return to happiness after you are derailed from it, but that also you drop back down even if your happiness is elevated by, say, winning the lottery.

'People get used to stuff,' says Michael Plant. 'Nothing feels very good or bad for very long. Births, deaths, promotions, demotions. Some people have advanced set point theory — whatever happens you come back to a set point.'

Given there is evidence for this set point we're always going to come back to, you have to ask: is there anything we can do to increase our happiness over the long term?

The first thing we need to do is to understand how the base-line is set in the first place. The problem is, this is incredibly complicated. It's not controversial to suppose there is a genetic contribution to it, since there's a genetic contribution to most things, as we've seen throughout this book. What's contro-versial are claims to have found a specific genetic element that supposedly makes people happier (or more prone to depres-sion). You can imagine what's at stake here if and when such genetic components are found. I can already see the marketing campaign for the happiness pills.

One such claim has sparked huge amounts of research over the last fifteen years.

The variant in question is found in the serotonin-transporter gene on chromosome 17. You'll have heard of serotonin, a chemical commonly associated with feelings of happiness. When there is more of it in the bloodstream, we tend to feel happier, which is why stopping it from being mopped up, by using a drug such as Prozac, can benefit people suffering from depression. The serotonin-transporter gene's job is to make a protein that carries serotonin away to be recycled.

As in most genes, there's a region of the serotonin trans-porter that controls how much of the gene product is made. It controls expression of the gene, which is like saying it's the tap regulating water flow into a bath. In 1994 geneticists discov-ered that this region, 5-HTTLPR, occurs in two forms, a long and a short form. People with the long form make more of the product than those with the short form, and more of the trans-porter product means less serotonin is left hanging around than in people with the short form. This discovery immediately had the neuroscientists and psychiatrists wondering: what would the behavioural effect of this difference be in people?

In 2003, a bombshell paper in the journal *Science* suggested an answer.[8] A small study of young men and women in Dunedin, New Zealand, found that people with the short form of the gene were more prone to depression, and even suicide, than people with the long form.

Hundreds of research papers have since attempted to delve into the effect of the 5-HTTLPR gene.[9] It's not clear cut, and with a condition as complex as depression, known to be influenced by multiple factors, you wouldn't expect the behaviour to map simply to a single gene. But the evidence was starting to suggest that people with the short form do show higher rates of depression. One analysis of data from 2574 teenagers enrolled in the US National Longitudinal Study of Adolescent Health allowed a comparison of genetic status with life satisfaction responses. Jan-Emmanuel De Neve, now based at the Saïd Business School and the University of Oxford, found that people with the long variant of 5-HTTLPR were happier. This, he proposed, could explain differences in baseline levels of happiness.[10]

But De Neve's own follow-up provided mixed results.[11] What this tells us, as Robert Plomin – the intelligence researcher we met in Chapter 1 – warned, is that you need huge data sets to find relationships between gene variants and complex traits. It's no good to sample the DNA of a few hundred or even a few thousand people for your study. You need DNA samples from hundreds of thousands, preferably millions, of people. Those data sets are only just becoming available. In 2016, De Neve was among the authors of one such study, which looked at the genetic influences on subjective well-being in 298,420 people.

Among the millions of differences in DNA, the team found three genetic variants (two on chromosome 5, and one on chromosome 20) that were associated with well-being. Together they account for just 0.9 per cent of differences in well-being between people. So on the one hand we have robustly discovered genetic elements that definitively contribute to well-being. On the other, the explanatory power of each variant is minute. Even adding together all the genetic influences, including the thousands of as-yet unknown variants, De Neve says, the environmental influence on happiness and depression will likely be greater than that caused by genetics.

These larger data sets have also muddied the water when it comes to 5-HTTLPR. In 2017 there was a major re-analysis of all the data on 5-HTTLPR, stress and depression. The authors, and there are dozens of them, examined thirty-one data sets containing information on 38,802 people. The conclusion convincingly deflates the idea that there is a link between the short 5-HTTLPR genotype and depression.[12]

What all this means is that the search for a 'happiness gene' is over. The search for the thousands of variants that together boost our happiness, however – that continues. The lead author of the 2017 meta-analysis is Robert Culverhouse, a professor at Washington University St Louis School of Medicine, in Missouri. A consensus has been reached in recent years, he says. For complex traits, such as those examined throughout this book, there are many hundreds of genetic variants that interact to influence the development of the trait: 'The impact of any particular genetic variant in isolation is modest.' This means we should not expect easy gene-based therapies, such as gene editing or a drug to target a single gene, and should be sceptical of any claims to the contrary. 'What seems to be a more immediate application of genetic findings for complex traits,' Culverhouse says, 'is personalised treatment based on a genetic profile.'

Though the effects are modest, it is still very much worth finding the variants associated with different traits. They will lead to insights into the mechanisms by which the traits develop, and could lead to new treatments. Remember, although the confirmed links between genetic variants and well-being only explain a tiny amount of the trait, that doesn't mean there aren't a large amount of other, undiscovered genetic links. There are. Genetics plays a major role in depression, and De Neve's work – using twin studies – shows that it accounts for about a third of the differences between people in well-being.

Interest in 5-HTTLPR won't go away. Another author of that 2017 paper is Kathryn Lester, a psychologist at the University of Sussex. One of the problems even with meta-analyses of well-being and depression is, she tells me, heterogeneity – unevenness – in the sampling. For example, some studies might measure individual well-being using self-reported assessments, while others use face-to-face interviews, and the two methods have different levels of accuracy. A meta-analysis tries to minimise this unevenness but is still vulnerable. 'So,' says Lester, 'I think the controversy regarding the interaction between 5-HTTLPR genotype and stress exposure on depression may very well continue.'

Lester, like Culverhouse, says that while we are quite a long way off gene-based treatments to improve our well-being, even this complex, emerging understanding of the genetics is promising. It will enable researchers to identify biological pathways in the brain and people who may be more or less responsive to existing treatment options. For now, however, it does seem like we can't move the baseline that is our initial setting.

*

Tim Harrower, Shirley Parsons's neurologist, says physical recovery from locked-in syndrome doesn't really happen: 'That's the sad part about it. It is generally very, very limited.' Generally, but not always. One neuroscientist I talked to suggested I get in touch with a woman called Kate Allatt. She is living proof that remarkable recovery does sometimes happen.

When Allatt was a schoolgirl in Sheffield, she says, she messed around a fair bit; she was a mischievous kid. But she was indignant when a careers counsellor at school advised her to aim only for factory work. 'That helped me,' she says. 'Now, if they say "you can't", I say "watch me".'

She says she has always been positive and determined. She grew up to become a fell runner, putting in 70 miles per week. She had three kids and a full-time job. But even those challenges were nothing compared to the one she faced on 7 February 2010, when she was thirty-nine.

Allatt was resting in bed with what doctors had thought was a migraine. That evening she managed to totter downstairs to her husband and asked him, 'What's happening to me?' – but what came out was slurred and unintelligible. She collapsed, as a massive brainstem stroke took hold.

When she came round, in Sheffield's Northern General Hospital, she said it was like waking up in her own coffin. She was in intensive care, completely paralysed; locked-in, with a machine breathing for her. Her husband was told she'd be better off dead. Allatt, completely conscious, was terrified the doctors would turn the life-support machine off but was unable to communicate her fears.

Eventually, she started communicating with friends and family by using eye-blinks. She was desperate to recover, and worked furiously to improve. 'My family were told to get on with their lives,' she told me. 'They wrote me off, they immediately lowered expectations before we'd even had a chance to try, and that pissed me off. I said you know what, I'm not taking that. I've got three kids at home, I'll try everything. I tried really hard, over and over and over and over again.'

Her entire focus was directed at trying to regain movement. After eight weeks in intensive care, she managed to move her right thumb two millimetres. She went from there. She managed to speak again. And incredibly, against the odds, she managed to walk out of hospital on her own (the video of this is on YouTube, and it's very moving). A year later, she did a one-mile run. Now, recovered, back with her family, she spends her time giving talks, and visiting people who are locked-in, helping them try to improve their situation and encouraging them with her story.

Parashkev Nachev, a neurologist at the faculty of brain sciences, University College London, is familiar with Allatt's case and how she recovered. I get in touch with him to find out how she did it. The brain, he explains, develops through self-organisation, and requires randomness and chance in order to achieve the complexity it needs. In this sense, it develops through trial and error. When it is damaged, it can either reorganise in order to regain lost function, or it can stay damaged. But reorganisation takes a lot of flexibility, in order to discover and test out new ways of doing things. 'This is where Kate comes in,' says Nachev. 'Her insane, superhuman drive meant she did a lot more "trialling" than others in her condition, indeed probably any patient I have ever come across, and so the brain was able to self-organise to a far greater extent than an ordinary human being.'

So at the end of the book we've finally met someone actually referred to as a superhuman. For her part, Allatt says that the key to improvement is trying: 'I've been saying that for years and the scientists are finally catching to my way of thinking.'

Allatt, like Will MacAskill, has found that helping others improves her own well-being.[13] When I ask her what her experience has told her about happiness, she tells me about a Finnish former model she's got to know. In 1995 Kati van der Hoeven was living her dream. She was twenty years old, an international model, living in Los Angeles. She was visiting her family in Finland when she had a massive stroke and became locked-in. That was twenty-two years ago. She is still paralysed, but now she is married, she lives in her own home, and she is happy – Allatt says Kati and her husband are the happiest people she's ever met.

Of course, Kati can't talk, so I contact her on Facebook and email. Her replies are so fast I think someone must be answering for her, but then she shows me a video of how she uses the computer. She has a little reflective dot on her forehead, and an infrared sensor on her computer picks up where she's looking. With this she can swiftly guide a cursor around a virtual keyboard.

She misses dancing, and of course she misses talking. But she says she no longer considers what happened to her a tragedy.[14] She is happier now than when she was a model. 'For me happiness is love, not just receiving it but also being able to give it, to share it,' she says. (Not just with her husband, Henning, but also with her dog, named Happy.) 'Secondly,' she says of happiness, 'I would say it's having a purpose in life and making a difference in other people's lives, as little that it may be.'

The Finns have a word, *sisu*, which means something like 'grit in the face of adversity', or 'courage against terrible odds'. It has a meaning of equanimity, like we saw in the parable of the farmer and the horse, and also of stress management. (Equanimity, the treatment of both good and bad events with the same detached response, is one of the Seven Factors of Enlightenment taught by the Buddha.)[15]

Many locked-in people get in touch with Kati. They read her blogs or see her on YouTube, or see her give a talk with Henning. The majority of them, she says, have not come to terms with what has happened. 'By this, I mean they do not look on the bright side and they don't get to learn about themselves.' What she wants to do is teach them *sisu*. Before her stroke, she says, all her actions were guided by societal rules; now they are mostly guided by emotions: love, compassion, kindness. 'My actions are not so much influenced by society but more so by what feels right or wrong.'

There are two things I take away from this. One is that the majority of people who contact her haven't made peace with their condition. For all I've been citing the research that shows most locked-in people are happy, it would be crazy to pretend it's not a crushing blow to come to terms with.

The other thing is that even those with natural *sisu* and positivity need a purpose. If you're going to suffer locked-in syndrome, it will help to have a positive outlook on life. 'I don't doubt that my happy disposition and inability to maintain upset or anger has helped me,' Shirley says. 'I think that it also helps that I'm very practical and realistic.' Shirley's case shows to what extent happiness doesn't require beauty, athleticism, fame or riches. What you experience from moment to moment, how you feel in yourself, doesn't come from these things. Nor even is the other main measure of happiness, the lifetime evaluation measure, irrevocably tied to these things we all chase. If you look back on a life and have spent twenty or thirty years of it locked-in, you might think, and you'd surely be justified, that you've been dealt a rough hand. You might, if you were Shirley, look back and say, well, shit happened, but I did some good things, I took those degrees, I improved my mind; I understood myself and the meaning

302

of life. Shipwrecked in my own body, I explored myself, and I found hidden treasure.

Nachev wonders what insight these stories about locked-in people gives us into the ordinary notion of happiness: 'Someone might profess to be happy in circumstances most of us would find unbearable. But I am not sure it tells us very much what happiness is for the rest of us.'

I think it does. For me, it drives home how the experiential, moment-to-moment measure of happiness is the one that is more tangible, and, I think, more important. It's the one we think of if someone asks if we're okay. Let's not over-complicate things. Happiness is simple. It's a bodily feeling, like being hot or cold, and we it know when we have it. When we want to position ourselves more grandly, or take a wide-angle view, we think about the life evaluation measure of happiness. As Plant said to me, 'We're often driven by the story of our lives rather than the moment-by-moment experience.' We know, from poems, songs and novels, that money doesn't make us happier. But we race along on a treadmill trying to get more. When Jefferson wrote about the right to 'the pursuit of happiness' in the declaration, I think he meant the pursuit of greater well-being in society as a whole, but it's become confused, in an individual-first world, with the personal life-evaluation measure of happiness. We think the pursuit of happiness means making sure we get that pay rise, but we've seen in this chapter that after a certain amount, fairly low in the middle-class scheme of things, extra money doesn't do much for our well-being. We think it means aiming for Kanye's car, or a celebrity lifestyle we think is cool, or a certain size house, or a particular school for our kids, and we fashion a story about how that will bring life satisfaction, and we are willing to endure commutes and long hours and tedious jobs – maybe even tedious partners – because we think they will help us reach this point of happiness. We may well get there, and that's great — but our moment-to-moment emotional happiness may not be altered, and we need to understand that and be ready for it or we'll always be bashing our heads against the wall. Perhaps an evolutionary viewpoint can help. It made good evolutionary sense hundreds of thousands of years ago to place great value on resources such as shelter and access to food and tools. People who had them acquired higher status. Now that desire for resources has been co-opted; it has become a runaway need for more and more. It's analogous to how obesity can be seen as an evolutionary trap – we evolved appetites at times when sweet and fatty food was far rarer than it is today (and when we had far more active lives), and our bodies can't cope with the vast amounts of cheap energy they have access to today.[16] Consumerism is a similar kind of evolutionary trap.

Surely the simplest change we can make in our lives, in our mindsets, is to remember that the moment-by-moment experience is what we should think of when we think about our happiness. Kanye might not be happy in his Maserati. Set your life up so as much as possible you are doing things that make you happy in the moment.

CONCLUSION

Individual citizens are internally plural: they have within them the full range of behavioural possibilities.

Zadie Smith

Man is a rope stretched between the animal and the Superman — a rope over an abyss.

Friedrich Nietzsche, *Thus Spoke Zarathustra*

Let me make it plain, you gotta make way for the Homo superior.

David Bowie, 'Oh! You Pretty Things'

A few days ago, cycling along the Mall towards Buckingham Palace, as it happens, I saw a child being admonished by his parents. 'Just who do you think you are?' the mother shouted. I cycled on and didn't hear any more (given where we were, maybe the next line was 'Do you think you're the blimmin' Queen?') but I thought about that 'who do you think you are' in the light of the people I've met in this book. Who do I think I am? I think I am part of a species with far greater potential than I realise in my day-to-day life.

When I read and marvel at accounts of remarkable survival or bravery, or admire great works of art, literature and science as the pinnacles of human achievement, I feel that the people who perform and achieve such feats are otherly. We masses are not superhuman, nor are most of the people we meet. What I've belatedly realised is that there is a thread connecting us to them. We can bask in the glory of these others of our species, because we have similar traits too.

Let's see. My memory is only average, but I could train it to be better, if I wanted. I very much doubt I could run the Badwater Ultramarathon, but I might be able to make it round a regular marathon, if I put my mind to it. I will never win a science Nobel Prize, but I did once have a paper published in *Science*.[1] I will never sing at the Royal Opera House, but at least now, after practising, I don't disgrace myself at karaoke.[2] To some extent, at least, there is potential within us all.

We've covered a diverse array of traits in our survey. Intelligence and bravery, singing and endurance, resilience and sleep, old age and happiness. But there is a unifying link between them: there are degrees of them, and are expressed in all of us. They are part of the colour palette that makes us human; perhaps they are the prime colours in it.[3] Hamlet's 'what a piece of work is man' speech comes to mind, but after praising humanity's capacity for reason and our infinite faculty, Hamlet sourly dismisses it: 'Man delights not me; no, nor woman neither'.[4] That's where we differ: at the end of our survey, I am far from down on humanity – Hamlet is on a downer, and I am on an upper.[5]

As we've seen, superhumans are just at the end of a spectrum of ability. What is controversial is how they got there. For several of the traits we've examined, it seems certain that the people I've met are the carriers of precious genetic cargo that have helped them achieve greatness. Intelligence, singing ability and longevity are the clearest examples. But even with these you need a favourable environment to facilitate their full development. Mozart wasn't born playing the piano; Magnus Carlsen had to learn the game of chess. With other traits — memory and endurance, for example — you can ratchet yourself to a high level regardless of your genetic deck.

We've seen throughout this book that how well you are able to do something depends on how you are raised, what you eat and what you experience, how you train and practise, as well as what you are born with. The environment nurtures, inter- acts with and modifies your genes. It does not overcome the influence of your genes — it works with them, turning them on and off, boosting some and suppressing others. There is no such thing as 'nature versus nurture'. It is never genes *or* environ- ment; it is always both things, together. The phrase persists in the popular imagination, an ancient meme well past its sell-by date, but it has long been dismissed by geneticists as simply wrong. There's another quote from *Hamlet* here that is apt: 'For use almost can change the stamp of nature.' The prince is trying to stop his mother sleeping with his uncle, saying that if she practises *not* going to Claudius's bed, her innate need to do so can be reduced. I marvel that Shakespeare put that 'almost' there. Training can *almost* trump innate ability, is my reading of this.[6]

Complex traits are influenced by thousands of genes. A gene confirmed to have an effect on intelligence, for example, will also play roles in many other traits. Genetic influence is smeared. It's not surprising, since intelligence and, say, memory, or language ability, or resilience, are just categories we've assigned to help describe and understand things about humans. These categories make messy reality manageable; they are not Platonic ideals. The technical name for this smearing effect in genetics is pleiotropy. Genes linked to intelligence will probably also influence memory, language ability, focus and happiness, and probably longevity and resilience. And vice versa. For example, one study of Vietnam War veterans found a link between intelligence and health, perhaps because low intelligence and poor health share genetic factors,[7] and an analysis of three studies of different twins showed that there was a small association between intelligence and lifespan – smarter people live longer – and that this link was genetic.[8]

This smearing and interrelatedness do not mean that genetics plays less of a part in influencing our athletic ability, or our intelligence, or our likelihood to live to one hundred: there is still a hefty genetic component. Except for a small number of mutations, such as the Mendelian diseases like cystic fibrosis we saw in Chapter 9, you can't draw a line from a particular gene to a particular behaviour or physiological effect.

So why the resistance to the idea of nature? Any number of best-selling books recently (I'm thinking of *Grit* by Angela Duckworth; *Peak* by Anders Ericsson and Robert Pool; *The Talent Code* by Daniel Coyle; *Outliers* by Malcolm Gladwell; *Talent is Overrated* by Geoff Colvin; and how's this for a wishful title: *The Genius in All of Us*, by David Shenk) emphasise the role of nurture over that of genetics and downplay the role of innate ability.

Partly it's because how we commonly think about genetics is wrong. The meme sets nature *against* nurture, and people think that nature is immutable and unmodifiable. It is not. Modifications can occur in any number of the traits we've encountered in this book. Genes turn on and off according to the demands put on them, and they are modified by training, diet and other environmental influences. The study of this kind of modification is called epigenetics.

The unwillingness to accept a genetic influence in what we do could be ideological. No one likes that idea that genes control our destiny. Well, no one is saying that; it's a straw man argument. Then there's the taint of history, and the horrors of the eugenics movements in the UK, Scandinavia, Canada and the US, and of course Germany. Yet that is politics, not science. If modern science tells us that there is a genetic component to high achievement in this or that trait, as we've seen in this book, then that's the way it is. Accept the evidence, and be empowered. For one thing, it will enable people to devote their resources to the right place. For another, once we understand more about the genetic influence on expertise then more people will be able to accomplish greater things through training. It's similar to the argument advanced for measuring IQ: by doing so we can better help those who traditionally have been neglected. David Card and Laura Giuliano at the University of California, Irvine, showed how the introduction of IQ screening in a large school district in Florida led to a large increase in the number of black, Hispanic, low-income and female students into programmes for gifted children[9] and Robert Plomin has argued that IQ testing can improve educational attainment for all children.[10] Incidentally, imagine if there was no genetic impact on intelligence. If differences in nurture alone determined intelligence, then everyone from deprived backgrounds would be less intelligent than those with advantageous backgrounds.

What does this mean for you as a reader setting out to become an expert in something, or for those of you who are parents or want to be parents, thinking about how to best guide your children? It means not setting unrealistic targets. It means trying different things — be Goldilocks about things — until you find something that suits you. If you find you have a knack for something, keep going. If you realise you've not got what it takes to pursue something as a career, stop. Do it because it's fun, not because you want to become an expert. Remember too that part of our genetic endowment is our drive. This was one of the biggest revelations for me in researching this book, discovering that the propensity to practise itself has a strong genetic element, as we saw in Chapter 6. I'd never thought about traits such as doggedness or grit having a genetic component. But they do. Don't beat yourself up if you don't reach the level of the people in this book: very few do.

I remember once asking my mother, a successful writer of books for children and young adults, why she bothered. I think we'd both just read a Philip Pullman novel. She's sold a lot of books, but by her own admission, she wasn't going to be a Pullman, or a Roald Dahl or a J.K. Rowling. She did it, she said, because she was happy to contribute a small amount. Meeting the superhumans in this book hasn't made me quake in terror of not being able to achieve what they've done; they've inspired me to put a little more in to what I do. Run, or at least do exercise, for the positive vibes it brings you. Prioritise your sleep as you would your safety crossing the road, or the health of your children – it will make you feel better, think better, and it will pay off in the long term, too. You don't have to put your mind to something so doggedly that you are able to race a Formula 1 car, or sail round the world, but a little thinking about how to live in the moment may improve your life and maybe even your happiness. I hope I won't ever find myself in a position of such horror as Carmen Tarleton or Alex Lewis, but lesser setbacks affect us all. We are resilient; we are equipped to get through them.

For me, this is enough, but I know that many people won't accept non-superhuman status. It's one of the reasons why those books I mentioned sell well, and why there is such intense genetic research going into areas such as intelligence, musical ability and longevity. Even sleeping. Remember the *DEC2* gene we saw in Chapter 10? People with a particular version of the gene can get by with fewer hours of sleep than the rest of us, and some scientists are quite open in saying that people will be genetically enhanced in the future.[11] The Harvard geneticist George Church talks about genetic variants that he says could potentially be edited into the human genome.[12]

The rise of gene-editing techniques such as Clustered Regularly Interspaced Short Palindromic Repeats (CRISPR) has generated huge excitement[13] (and anxiety). Research groups in China[14] and the US[15] have already tried (with mixed success) to use the technique to modify human embryos. Even when the logistics have been worked out, and its safety demonstrated, the pleiotropic nature of our genes points to how hard it's going to be to engineer significant changes into humanity. Intelligence, longevity, musicality and personality are far more genetically complex than we once thought. That said, people will try. Genetic engineering has resulted in intelligence and lifespan boosts in animals. Rats have been made smarter by modifying a single gene, *NR2B*, which improved their learning and memory skills.[16] This gene seems to allow neurons to communicate with each other for longer. In the worm *Caenorhabditis elegans* tweaks to two genetic pathways linked to ageing resulted in a five-fold increase in lifespan.[17]

Here's a question. Modern humans evolved some 200,000 years ago. Are we 'more' human now because some members of our species have done great things? No – because just as there is a thread connecting us contemporary people to the superhumans in this book, so is there a thread going back to prehistoric people. But look at us. We're taller, healthier, smarter; we live longer, we achieve more. It's not that we're more human, it's that we've achieved more of our potential. How much more is there?

There's a lot more. Here's just a glimpse.

In 2016 and 2017 there were unprecedented breakthroughs in the understanding of the ancient game of Go. What had happened was an artificial intelligence (AI) called AlphaGo competed against the world's top human players, and crushed them. It played moves that had never been seen before, in the game's 3000-year history. But the best humans in the world upped their game. Lee Sedol, of South Korea, and Ke Jie, of China, changed and improved the way they play because of what the AI had shown them. 'After my match against AlphaGo, I fundamentally reconsidered the game, and now I can see that this reflection has helped me greatly,' said Jie. 'Although I lost, I discovered that the possibilities of Go are immense and that the game has continued to progress.'[18] Jie then went on a twenty-two game winning streak.

Demis Hassabis is co-founder of Google DeepMind, the London-based lab that developed AlphaGo, and its even more impressive successor, AlphaZero.[19] He says the response of Jie and Sedol shows what AI can do for humanity. We fear that AI will take our jobs, but this is misplaced. AI shows us who we can be. 'Human ingenuity augmented by AI will unlock our true potential,' Hassabis says.

I can imagine how AI could augment and improve several of the traits we've covered in this book. Intelligence and creativity, certainly. Also memory, language and focus. I'm sure breakthroughs in medical science aided by AI will influence our longevity and resilience too. Will this mean we become post-human? Will it widen the chasm between the haves and the have-nots?

Even without AI, the people I've met over the course of

writing the book have left me excited about the potential of the human species. There's a lot left in us. This is something else that's been driving me to write this book.

Anders Sandberg, based at Future of Humanity Institute in Oxford, is among other things interested in measuring the future potential of humanity. We face non-negligible existential threats. There's climate change, synthetic biology, nuclear war and (ironically, since I've just extolled its virtues) artificial intelligence. One way of measuring the risk we face is to calculate what philosophers call the 'size of the future', in other words the number of potential future lives that are possible. Sandberg has done this, and the numbers are beyond astronomical. Current existential threats to humanity risk between 800 billion and 3.92×10^{100} future human lives.[20] We should get our act together, he says, for the sake of these future lives, and for the sake of our potential.

In this book we've explored examples of the best of us. We started by looking at the ways we achieve more than chimpanzees, and the traits in this book – our culture, our humanity – go a long way to explaining that. The other thing, of course, that we do at a higher level than chimps is cooperate. This quality is needed more than ever. I have marvelled at the richness of the human species and thrilled at our possibilities. We must harness these traits, the current extremes of human potential – and use them solve the problems facing our species.

ACKNOWLEDGEMENTS

Writing this book has been immensely enjoyable and fulfilling, and all the credit for that goes to the people I've met and interviewed along the way, both scientists and those who are superhuman subjects in the chapters. There are too many of you to mention by name here – you are named in the text – but thank you for your generosity of time in meeting and talking to me about your life and work. I had no idea quite how inspirational it would be to meet you and hear your stories.

New Scientist is a wonderfully stimulating organisation, and being part of the magazine for more than ten years has hugely informed how I think. Many thanks to my colleagues for their part in creating an atmosphere of inquiry and imagination, and thanks too to my bosses for allowing me the flexibility I needed to write this book.

Several people read and provided comments on various parts of the book: Celeste Biever, Jessica Hamzelou, Laura Gallagher, Simon Fisher, Marie-Christine Nizzi, Stuart Ritchie, Miriam Mosing and Jakke Tamminen. Many thanks for taking the time; any errors, of course, are mine. For advice and encouragement along the way, thanks to fellow-writers Helen Thomson, Gaia

Vince, Jo Marchant and Vic James. Jakke Tamminen and David Morgan very kindly wired me up to an EEG machine and let me sleep in their lab. Meeting all the people in the book took lots of organisation, which couldn't have been done without help and great good will from many, including: the residents and staff at the Royal Hospital Chelsea; the team at Williams Grand Prix Engineering; the Royal Opera House; Brigham and Women's Hospital, Boston; Debora Price of the British Society of Gerontology; and Tim Harrower of the Royal Devon and Exeter Foundation Trust Hospital.

The editorial team at Little, Brown – Tim Whiting and Nithya Rae – and copy-editor Steve Gove, as well as Ben Loehnen at Simon & Schuster, have been fantastic. I'd particularly like to thank my awesome and tireless agent, Patrick Walsh, who was instrumental in turning an initial idea into a book.

Thanks to my amazing mum, who raised me to the clacking sound of a typewriter – she showed me I could be a writer too. Thanks to the rest of my brilliantly supportive family and extended family – my dad, my sister Gemma, and all the in-laws (special shout out to Ros – lifesaver), and my children, Molly and Iris.

People say that writing a book is like having a baby, but while I was working on this book, my partner Laura really was gestating a baby. I'd like to thank her not only for tolerating my occasional physical and mental absence during this process, but also for the huge amount of love and support she gave me along the way. As the first reader I write for, she has massively improved what I do, but more than that, she's improved my life, too. This book is for her.

CREDITS

Memory

Pliny, *The historie of the world*: commonly called *The naturall historie of C. Plinius Secundus*. Translated by Philemon Holland Doctor of Physicke (https://quod.lib.umich.edu/e/eebo/A09763.0001.001/1:48. 24?rgn=div2;view=fulltext)

Sartre, *Nausea*, translated by Robert Baldick (London: Penguin, 2000)

Wilson, *Plaques and Tangles* (London: Faber, 2015)

Language

Wallace, *Infinite Jest* (London: Abacus, 1997)

Focus

Yamamoto, *Hagakure*, translated by Alexander Bennett (Vermont: Tuttle Publishing, 2014)

Running

Murakami, *What I Talk About When I Talk About Running* (London: Vintage, 2009)

Sleeping

Tranströmer, 'Nocturne', translated by Robert Bly, *The Half-Finished Heaven* (Minnesota: Graywolf Press, 2017)

Happiness

Smith, *Hotel World* (London: Penguin, 2002)

Conclusion

Smith, 'On Optimism and Despair', *New York Review of Books*, 22 December 2016

Nietzsche, *Thus Spoke Zarathustra*, translated by R. J. Hollingdale (London: Penguin, 1961)

Bowie, 'Oh! You Pretty Things', on *Hunky Dory* © RCA 1971

REFERENCES

INTRODUCTION

1 *Current Biology*, DOI: 10.1016/j.cub.2010.11.024
2 *Current Biology*, DOI: 10.1016/j.cub.2006.12.042
3 *Proceedings of the National Academy of Sciences*, DOI: 10.1073/pnas.07026 24104
4 *Journal of Human Evolution*, doi.org/smp
5 *Scientific Reports*, DOI: 10.1038/srep22219
6 https://www. newscientist.com/article/mg23130890-600-metaphysics-special-where-do-good-and-evil-come-from/

1   INTELLIGENCE

1 *Frontiers in Psychology*, DOI: 10.3389/fpsyg.2014.00878
2 http://www.telegraph.co.uk/men/the-filter/football-mad-mobbed-by-girls-and-easily-bored-meet-magnus-carlse/
3 *Psychology and Aging*, DOI:10.1037/0882-7974.22.2.291
4 *Journal of Intelligence*, DOI: j.intell.2017.01.013
5 *Intelligence*, 45 (2014), 81–103 http://dx.doi.org/10.1016/j.intell.2013.12.001
6 *British Medical Journal*, DOI: 10.1136/bmj.j2708

7   For more on this see: http://slatestarcodex.com/2017/09/27/against-individual-iq-worries/

8   *Molecular Psychiatry*, DOI: 10.1038/mp.2014.105

9   https://www.theatlantic.com/health/archive/2014/01/the-dark-side-of-emotional-intelligence/282720/

10 *Journal of Applied Psychology*, DOI: 10.1037/a0037681)

11 http://www.teds.ac.uk

12 Plomin et al., 'Nature, nurture, and cognitive development from 1 to 16 years: a parent-offspring adoption study'. *Psychological Science*, 8 (1997), pp. 442–7

13 *Intelligence*, DOI: j.intell.2014.11.005

14 *Nature Genetics*, DOI: 10.1038/ng.3869

15 *Psychological Science*, DOI: 10.1111/j.1467-9280.2007.02007.x

16 *Nature Communications*, DOI: 10.1038/ncomms2374

17 *Psychological Science*, DOI: 10.1111/j.1467-9280.2008.02175.x.

18 *Trends in Cognitive Sciences*, DOI:10.1016/j.tics.2016.05.010

19 https://www.newscientist.com/article/mg23431260-200-how-to-daydream-your-way-to-better-learning-and-concentration/

20 *Developmental Psychology*, DOI: 10.1037/a0015864

2   MEMORY

1   Thanks to Felipe De Brigard for pointing this out; see *Synthese*, DOI: 10.1007/s11229-013-0247-7

2   https://stuff.mit.edu/afs/sipb/contrib/pi/pi-billion.txt

3   *Neurocase*, DOI: 10.1080/13554790701844945

4   http://www.worldmemorychampionships.com

5   *Neuron*, DOI: 10.1016/j.neuron.2017.02.003

6   *Neurocase*, DOI: 10.1080/13554790500473680

7   *Frontiers in Psychology*, DOI: 10.3389/fpsyg.2015.02017

8   *Memory*, DOI: 10.1080/09658211.2015.1061011

9   https://www.theguardian.com/science/2017/feb/08/total-recall-the-people-who-never-forget

10 PNAS, DOI: 10.1073/pnas.1314373110

11 *Nature*, DOI: 10.1038/35021052

12 *Psychological Science*, DOI: 10.1111/j.1467-9280.2008.02245.x

13 *Brain Structure and Function*, DOI: 10.1007/s00429-015-1145-1

14 *Synthese*, DOI: 10.1007/s11229-013-0247-7

15 *Psychology and Aging*, DOI: 10.1037/pag0000133

3 LANGUAGE

1 http://rosettaproject.org/blog/02013/mar/28/new-estimates-on-rate-of-language-loss/

2 http://www.polyglotassociation.org/

3 *Brain and Language*, DOI: 10.1016/S0093-934X(03)00360-2

4 *NeuroImage*, DOI:10.1016/j.neuroimage.2012.06.043

5 *NeuroImage*, DOI: 10.1016/j.neuroimage.2015.10.020

6 *Behavioural Neurology*, DOI: 10.1155/2014/808137

7 *PLoS ONE*, DOI: 10.1371/journal.pone.0094842

8 *Psychological Science*, DOI: 10.1177/0956797611432178

9 *Trends in Cognitive Sciences*, DOI: 10.1016/j.tics.2016.08.004

10 *Nature Genetics*, DOI: 10.1038/ng0298-168

11 *Nature*, DOI: 10.1038/35097076

12 *Proceedings of the National Academy of Sciences*, vol. 92, pp. 930–3

13 *Trends in Genetics*, DOI: 10.1016/j.tig.2017.07.002

14 *Nature*, DOI: 10.1038/nature01025

15 *Current Biology*, DOI: 10.1016/j.cub.2008.01.060

16 *Proceedings of the National Academy of Sciences*, DOI:10.1073/PNAS.1414542111

17 *Journal of Neuroscience*, DOI: 10.1523/JNEUROSCI.4706-14.2015

18 This is true when we learn a second language using traditional teaching methods, rather than when picking a language up by immersion. But what about when we learn our first language? We acquire complex grammatical rules and apply them successfully without being explicitly aware of the rules.

4 FOCUS

1 http://www.thedrive.com/start-finish/11544/jacques-villeneuve-says-lance-stroll-might-be-worst-rookie-in-f1-history

2 http://www.williamsf1.com/racing/news/azerbaijan-grand-prix-2017

3 *Cerebral Cortex*, DOI: 10.1093/cercor/bhw214

4 *Nature Reviews Neuroscience*, DOI: 10.1038/nrn3916

5 *Pain*, DOI: 10.1016/j.pain.2010.10.006

6 *Emotion*, DOI: 10.1037/a0018334

7 *Neuroscience of Consciousness*, DOI: 10.1093/nc/niw007

8 *Proceedings of the National Academy of Sciences*, DOI: 10.1073/pnas. 0707678104

9 One note of caution here: not everyone enjoys the positive benefits of meditation or mindfulness. Two UK-based researchers, Miguel Farias at Coventry University and Catherine Wikholm at the University of Surrey, have written about the potential unexpected consequences in *The Buddha Pill: Can meditation change you?* (London: Watkins Publishing, 2015).

10 *Frontiers in Psychology*, DOI: 10.3389/fpsyg.2014.01220

11 *Frontiers in Psychology*, DOI: 10.3389/fpsyg.2017.00647

5 BRAVERY

1 https://www.gov.uk/government/news/ied-search-teams-honoured-with-new-badge

2 http://www.bbc.co.uk/news/uk-england-25354632

3 *Nature Neuroscience*, DOI: 10.1038/nn2032

4 *Nature Neuroscience*, DOI: 10.1038/nn.3323

5 http://abcnews.go.com/US/hero-mom-describes-chased-off-carjackers-gas-station/story?id=36496307

6 *Neuron*, DOI: 10.1016/j.neuron.2010.06.009

6 SINGING

1 Anders Ericsson, *Peak: Secrets from the New Science of Expertise* (Eamon Dolan/Houghton Mifflin, 2016) p xviii

2 *Psychological Review*, DOI: 10.1037//0033-295X.100.3.363

3 *British Journal of Sports Medicine*, DOI: 10.1136/bjsports-2012-091767

4 Ericsson, *Peak: Secrets from the New Science of Expertise*, p 110.

5 *Psychological Bulletin*, DOI: 10.1037/bul0000033

6 *Intelligence*, DOI: 10.1016/j.intell.2013.04.001

7 *Frontiers in Psychology*, DOI: 10.3389/fpsyg.2014.00646

REFERENCES

8 https://alanwilliams123.wordpress.com/2015/06/30/northern-voices-opera-project-the-survey/ and https://www.thetimes.co.uk/article/opera-the-arsonists-to-be-sung-in-yorkshire-accent-alan-edward-williams-ian-mcmillan-39xfpkhjc

9 *Psychological Science*, DOI: 10.1177/0956797614541990

10 I'd always thought it was extraordinarily dismissive to consider that the creative songwriting genius of John Lennon and Paul McCartney was something you could simply practise towards, and then I read this comment by McCartney: 'There are a lot of bands that were out in Hamburg who put in 10,000 hours and didn't make it, so it's not a cast-iron theory,' he told *Q* magazine. 'I don't think it's a rule that if you do that amount of work, you're going to be as successful as the Beatles.' (From: http://www.cbc.ca/news/entertainment/interview-paul-mccartney-heads-to-canada-1.942764). In any case, a thorough analysis of the Beatles' Hamburg days found they performed together for a total of only about 1100 hours (from Mark Lewisohn, *Tune In*, New York: Crown Archetype, 2013).

11 *Psychonomic Bulletin and Review,* DOI 10.3758/s13423-014-0671-9

12 http://www.telegraph.co.uk/opera/what-to-see/written-skin-one-operatic-masterpieces-time-review/

13 http://www.lemonde.fr/culture/article/2012/07/09/written-on-skin-le-meilleur-opera-ecrit-depuis-vingt-ans_1731145_3246.html

14 http://www.japantimes.co.jp/life/2016/05/28/lifestyle/whispers-asmr-softly-rising-japan

15 Angela Duckworth, *Grit – why passion and resilience are the secrets to success*. Vermilion, 2017.

16 *Journal of Personality and Social Psychology*, DOI: 10.1037/0022-3514.92.6.1087

17 *Journal of Medical Genetics*, DOI: 10.1136/jmedgenet-2012-101209

18 http://jmg.bmj.com/content/45/7/451

19 *Scientific Reports*, DOI: 10.1038/srep39707

20 'Savant Syndrome: A Compelling Case for Innate Talent', DOI: 10.1093/acprof:oso/9780199794003.003.0007

21 *Frontiers in Psychology*, DOI: 10.3389/fpsyg.2014.00658

22 *Proceedings of the National Academy of Sciences*, DOI: 10.1073/pnas.1408777111

23 DOI: 10.1080/02640414.2016.1265662

24 *Developmental Review*, DOI: 10.1006/drev.1999.0504
25 *Perspectives on Psychological Science*, DOI: 10.1177/1745691616635600
26 Email correspondence with the author.

7 RUNNING

1 http://www.runnersworld.com/elite-runners/dean-karzes-runs-350-miles
2 http://www.dailymail.co.uk/news/article-3588109/Super-human-marathon-runner-Dean-Karnazes-jog-350-miles-without-stopping-thanks-rare-genetic-condition.html
3 http://www.bbc.co.uk/news/magazine-17600061
4 http://barefootrunning.fas.harvard.edu
5 *Journal of Sport and Health Science*, DOI: 10.1016/j.jshs.2014.03.009
6 http://www.menshealth.com/fitness/the-men-who-live-forever
7 http://running.competitor.com/2015/04/news/tarahumara-running-tribe-featured-in-a-new-documentary_125766
8 http://www.latinospost.com/articles/77223/20151102/tarahumara-athletes-win-world-indigenous-games.htm
9 *American Journal of Human Biology*, DOI: 10.1002/ajhb.22607
10 *American Journal of Human Biology*, DOI: 10.1002/ajhb.22239
11 https://www.nytimes.com/2015/03/07/sports/caballo-blanco-ultramarathon-is-canceled-over-threat-of-drug-violence.html
12 http://www.aljazeera.com/indepth/features/2016/01/running-lives-mexico-teenage-raramuri-160127090310518.html
13 https://www.theguardian.com/lifeandstyle/2015/mar/31/japanese-monks-mount-hiei-1000-marathons-1000-days
14 Thanks for the input, Celeste Biever.

8 LONGEVITY

1 Children learn from reading fairy tales that you have to be careful what you wish for from a genie. From fairy tales, or from science fiction. I remember reading Fredric Brown's short story, 'Great Lost Discoveries: Immortality', when I was a kid. The hero has invented an immortality pill, but is unsure about whether to use it. Finally, in

hospital, on his deathbed, his fear of oblivion overcomes his fear of for ever, and he pops the pill. He slips into a coma . . . and lives for ever. Eventually the hospital realise what's happened and, short of bed space, bury him.

2 http://www.un.org/esa/population/publications/worldageing 19502050/pdf/90chapteriv.pdf

3 *Human Genetics*: https://www.ncbi.nlm.nih.gov/pubmed/8786073/

4 *Human Genetics*, DOI: 10.1007/s00439-006-0144-y

5 http://www.nytimes.com/1997/08/05/world/jeanne-calment-world-s-elder-dies-at-122.html

6 http://web.archive.org/web/20010113103900/http://entomology. ucdavis.edu/courses/hde19/lecture3.html

7 *Journal of the American Geriatrics Society*, DOI: 10.1111/j.1532-5415. 2011.03498.x

8 Dong, X., Milholland, B. and Vijg, J., 'Evidence for a limit to human lifespan' *Nature* 538 (2016), pp. 257–9

9 *Nature*, DOI: 10.1038/nature22784

10 *Nursing Open*, DOI: 10.1002/nop2.44

11 http://www.dailymail.co.uk/femail/article-1008681/Alcohol-cigarettes-chocolates-sweets--The-secrets-long-life.html

12 *Nature Genetics*, DOI: 10.1038/ng0194-29

13 *Rejuvenation Research*, DOI: 10.1089/rej.2014.1605

14 *Neurobiology of Aging*, DOI: 10.1016/j.neurobiolaging.2012.08.019

15 *Nature Reviews Genetics*, DOI: 10.1038/nrg1871

16 *International Journal of Epidemiology*, 2017, 1–11, DOI: 10.1093/ije/ dyx053

17 *Journal of Gerontology*, DOI: 10.1093/gerona/glr223

18 *PLoS ONE*, DOI: 10.1371/journal.pone.0029848

19 *Journals of Gerontology Series A: Biological Sciences and Medical Sciences*, 15 March 2017, DOI: 10.1093/gerona/glx027

20 *Human Molecular Genetics*, DOI: 10.1093/hmg/ddu139

21 *Journals of Gerontology Series A: Biological Sciences and Medical Sciences*, DOI: 10.1093/gerona/glx053

22 This work, shared with me by Nir Barzilai, is currently in press in a cardiovascular journal.

23 *Frontiers in Genetics*, DOI: 10.3389/fgene.2011.00090

24 *Health Affairs (Millwood)*, DOI: 10.1377/hlthaff.2013.0052

25 *Neuroscientist*, DOI: 10.1177/107385840000600114

26 *Nature*, DOI: 10.1038/nature10357

27 *Cell*, DOI: 10.1016/j.cell.2013.04.015

28 *Nature Medicine*, DOI: 10.1038/nm.3898

29 http://bioviva-science.com/blog/one-year-anniversary-of-biovivas-gene-therapy-against-human-aging

30 *Proceedings of the National Academy of Sciences*, DOI: 10.1073/pnas.0906191106

31 http://onlinelibrary.wiley.com/doi/10.1002/emmm.201200245/full

32 You can see him here: https://www.youtube.com/watch?v=IEAejwYBilE

33 *Nature*, DOI: 10.1038/nature11432

34 Mattison et al., *Nature Communications* 8 (2017), DOI: 10.1038/ncomms14063

35 https://www.newscientist.com/article/mg22429894-000-everyday-drugs-could-give-extra-years-of-life/

9 RESILIENCE

1 http://nymag.com/health/bestdoctors/2014/steve-crohn-aids-2014-6/

2 *Nature Biotechnology*, DOI: 10.1038/nbt.3514

3 https://www.newscientist.com/article/mg21428683-200-watching-surgeons-expand-a-babys-skull/

4 *American Psychologist*, https://www.ncbi.nlm.nih.gov/pubmed/11315249

10 SLEEPING

1 *Scientific Reports*, DOI: 10.1038/srep17159

2 *Sleep*, DOI: 10.1016/j.slsci.2016.01.003

3 https://www.researchgate.net/scientific-contributions/33359804_Allan_Rechtschaffen

4 https://www.sciencealert.com/watch-here-s-what-happened-when-a-teenager-stayed-awake-for-11-days-straight

5 *Proceedings of the National Academy of Sciences*, DOI: 10.1073/pnas.0305404101

6 *Nature*, DOI: 10.1038/nature02663

REFERENCES

7 Otto Loewi, 'An Autobiographical Sketch', *Perspectives in Biological Medicine* 4 (1960), pp. 1–25
8 *Sleep*, DOI: 10.5665/sleep.1974
9 *International Journal of Dream Research*, DOI: 10.11588/ijodr.2009.2.142
10 https://www.newscientist.com/article/mg23331130-400-heal-yourself-from-inside-your-dreams/
11 *Imagination. Cognition and Personality*, DOI: 10.2190/IC.31.3.f
12 http://www.technologyreview.com/view/424608/extra-sleep-boosts-basketball-players-prowess/
13 *Sleep*, DOI: 10.5665/SLEEP.1132
14 https://www.theguardian.com/us-news/2017/may/31/what-is-covfefe-donald-trump-baffles-twitter-post
15 *Sleep Health*, DOI: 10.1016/j.sleh.2014.12.010
16 http://www.neurologyadvisor.com/sleep-2017/cvd-and-stroke-mortality-risks-linked-to-sleep-duration/article/668280/
17 http://www.hellomagazine.com/profiles/marthastewart/
18 *Trends in Genetics*, DOI: 10.1016/j.tig.2012.08.002
19 *Science*, DOI: 10.1126/science.1174443
20 *Journal of Sleep Research*, 10, pp. 173–9
21 *Science*, DOI: 10.1126/science.1180962
22 *Science*, DOI: 10.1126/science.1241224
23 *Brain*, DOI: 10.1093/brain/awx148

11 HAPPINESS

1 http://www.npr.org/sections/health-shots/2015/01/09/376084137/trapped-in-his-body-for-12-years-a-man-breaks-free
2 *Consciousness and Cognition*, DOI: 10.1016/j.concog.2011.10.010
3 *BMJ Open*, DOI: 10.1136/bmjopen-2010-000039
4 Richard Easterlin, 'Does Economic Growth Improve the Human Lot? Some Empirical Evidence' (pdf). In Paul A. David, Melvin W. Reder. *Nations and Households in Economic Growth: Essays in Honor of Moses Abramovitz* (New York: Academic Press, 1974)
5 http://www.nber.org/papers/w14282
6 *Proceedings of the National Academy of Sciences*, DOI: 10.1073/pnas.1011492107
7 http://www.latinamericanstudies.org/cold-war/sovietsbomb.htm

8   *Science*, DOI: 10.1126/science.1083968
9   *Molecular Psychiatry*, DOI: 10.1038/sj.mp.4001789
10 *Journal of Human Genetics*, DOI: 10.1038/jhg.2011.39
11 *Journal of Neuroscience, Psychology, and Economics*, DOI: 10.1037/a0030292
12 *Molecular Psychiatry*, DOI: 10.1038/mp.2017.44
13 http://www.huffingtonpost.co.uk/kate-allatt/wellbeing-volunteering_b_15425654.html
14 http://www.huffingtonpost.co.uk/author/henning-van-der-hoeven
15 I read this in James Kingsland, *Siddhartha's Brain* (London: Robinson, 2016).
16 *Nature Genetics*, DOI: 10.1038/ng.3620

CONCLUSION

1   *Science*, DOI: 10.1126/science.1064815 – this achievement was entirely due to the skill and generosity of the first author.
2   Actually this isn't true: I do disgrace myself.
3   Of course, in this book we have only looked at abilities and traits, not at emotions, such as love and sadness, which are also key to the human condition.
4   I can never read those lines without seeing Richard E. Grant magnificently spitting the words at the end of *Withnail and I*.
5   It's been claimed that in Hamlet, Shakespeare was describing the symptoms of what we now call bipolar syndrome. See: https://www.newscientist.com/article/mg22229654-900-shakespeare-the-godfather-of-modern-medicine/
6   It goes without saying that Shakespeare had an unprecedented and unsurpassed insight into human nature; what's interesting from my point of view and what makes him so quotable, five hundred years later, is that he seems to have intuitively understood the role of nature, of genetics, even if sometimes he over-eggs the pudding. Here's an example. Of wicked Caliban, Shakespeare has Prospero say he is 'A devil, a born devil, on whose nature Nurture can never stick.' Would Caliban always turn out evil? Is Prospero asserting that to salve his conscience? Or does Shakespeare – do we – really believe that someone can be *born* evil? Might Hitler have ended up

merely a grumpy old farmer if not for certain catastrophic turns of fate?

7   *Intelligence*, DOI: 10.1016/j.intell.2009.03.008

8   *International Journal of Epidemiology*, DOI: 10.1093/ije/dyv112

9   *Proceedings of the National Academy of Sciences*, DOI: 10.1073/pnas.1605043113

10 Kathryn Asbury and Robert Plomin, *G is for Genes: The Impact of Genetics on Education and Achievement* (Oxford: Wiley-Blackwell, 2013)

11 https://www.scientificamerican.com/article/improving-humans-with-customized-genes-sparks-debate-among-scientists1/

12 https://ipscell.com/2015/03/georgechurchinterview/ and https://ipscell.com/2016/04/new-chat-with-george-church-on-crispring-people-zika-weapons-more/

13 For example: https://www.newscientist.com/article/mg22830500-500-will-crispr-gene-editing-technology-lead-to-designer-babies/

14 *Molecular Genetics and Genomics*, DOI: 10.1007/s00438-017-1299-z

15 *bioRxiv*, DOI: 10.1101/181255

16 Public Library of Science, DOI:10.1371/journal.pone.0007486

17 *Cell Reports*, DOI: 10.1016/j.celrep.2013.11.018

18 Quoted in this tweet from Demis Hassabis: https://twitter.com/demishassabis/status/884915065715085312

19 https://en.chessbase.com/post/the-future-is-here-alphazero-learns-chess

20 These figures were presented at a public lecture at New Scientist Live, September 2017.

INDEX

INDEX

INDEX

INDEX

emotion 152
  'natural' singers 151–2
  practice and 146, 151
  speech and 150
  stamina requirement 144, 150
  successful composers 165
  vocal cords 149–50, 157, 158
  warming-up 157
optimism 239
orangutans 7–8
Ord, Toby 289
out-of-body experiences 170
overtraining 165
oxytocin 136, 137

Padron, Angie 135–6, 137
pain 107, 118, 134, 138
panic attacks 131, 132
parable of the Chinese farmer and
  the horse 285
Paralympics 140, 141
Parrish, Elizabeth 220, 221
Parsons, Shirley 279–80, 281–2,
  283, 284, 287–8, 294, 302
Patel, Aniruddh 166
Patihis, Lawrence 56, 57, 58–9
pattern perception 16
Pavarotti, Luciano 151
Paxton, Bill 241–2, 243
Penfield, Wilder 57
Per2 gene 274
periodic table of elements 264
Perls, Thomas 213, 214, 215, 217
perseverance 159
personality traits, alteration 66
Petkov, Plamen 125–6, 127, 129
Pfeiffer syndrome 243–4
Pheidippides 171–2, 175
Philosopher's Stone 197
pi, recounting of 38–9, 40–1,
  42–4, 45, 46
Pinto, Nicole 173, 174, 190
Pistorius, Martin 283–4
pitch-production accuracy tests 161
Plant, Michael 290, 294–5, 303

pleiotropy 307–8, 311
Pliny the Elder 37
Plomin, Robert 24–6, 33, 296, 309
police officers 118
polyglots 42, 67–72
  brain chemistry 72–3, 74–5
  genetic component 71
  hyperpolyglots 69, 72, 85
  introverted polyglots 76, 78
  lone studiers 70, 76, 83
  moral thinking, changes in 78–80
  and personality changes 75, 77
  social learners 70, 76–7, 83
  and thought pattern changes 76
Pomohac, Bohdan 228, 230, 231,
  232
Pool, Robert 308
positivity 207, 210, 238, 239, 302
Posner, Michael 107, 110
post-traumatic stress disorder (PTSD)
  116, 126, 128, 131, 132–3
  incidence of 131, 233
  symptoms 131, 232
  treatment 132–3
potential, human 184, 311, 312–13
practice
  deliberate practice 11, 12, 15, 46,
    148, 166
  genetic propensity to practise 154,
    155, 160, 174, 310
  10,000 hours of practice theory 11,
    146, 148, 154, 163
  and variance in performance
    outcomes 149
precuneus 267
prefrontal cortex 86, 88, 107, 262
Price, Deborah 220
Price, Jill 49, 50
problem-solving 30
procedural learning 86, 87
Profile of Mood States Test 109
Prozac 295
psychologist's fallacy 283
Ptacek, Louis 275–6
public speaking 137

341

INDEX

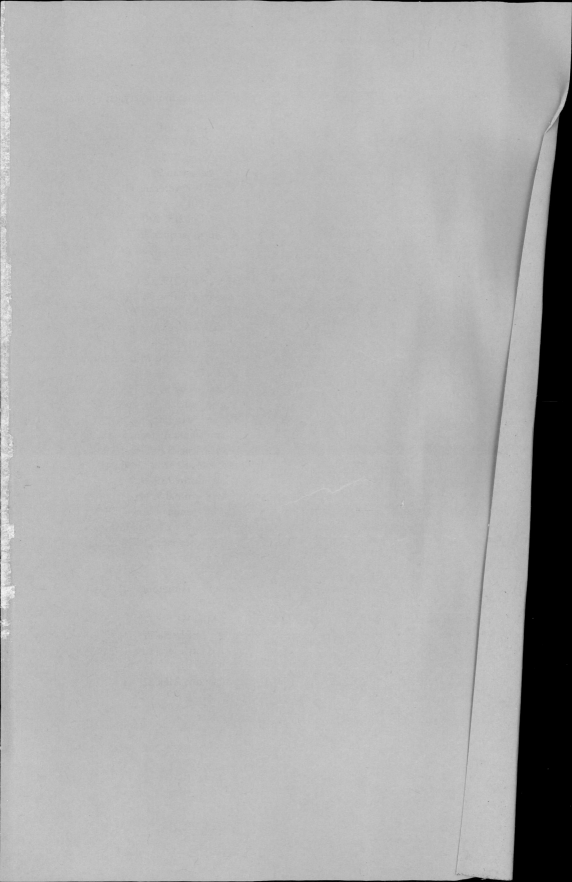